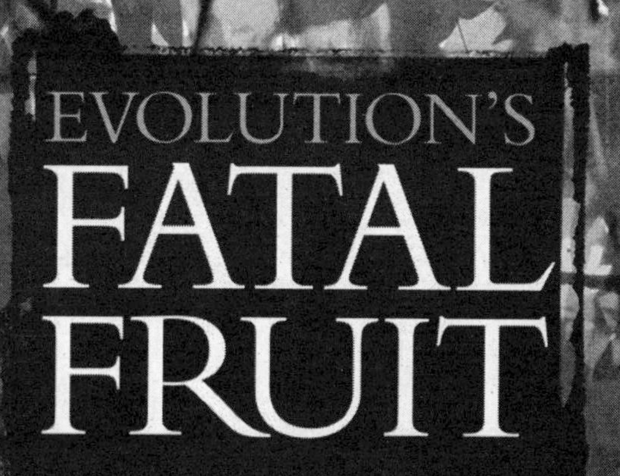

EVOLUTION'S FATAL FRUIT

How Darwin's Tree of Life Brought Death to Millions

Tom DeRosa

Foreword by D. James Kennedy, Ph.D.

Evolution's Fatal Fruit: How Darwin's Tree of Life Brought Death to Millions

By Tom DeRosa, Executive Director of the Creation Studies Institute

Published by Coral Ridge Ministries
Printed in the United States of America

Design by Roark Creative: www.roarkcreative.com

Coral Ridge Ministries
Post Office Box 40
Fort Lauderdale, Florida 33302
1-800-988-7884
letters@coralridge.org
www.coralridge.org

Creation Studies Institute
1001 West Cypress Creek Road, Suite 220
Fort Lauderdale, FL 33309
1-800-882-0278
info@creationstudies.org
www.creationstudies.org

CONTENTS

To Dr. Henry Morris,
founder of the modern creation movement,
and a man who will always be remembered for his strong
Christian witness and bold leadership in the long
war against evolution.

FOREWORD

The radio talk-show host sounded skeptical—even incredulous. "So to sum up," she asked me, "you see a direct connection between the scientific view of evolution and just about every instance of genocide in the twentieth century?"

I was talking to Terry Gross, the host of *Fresh Air*, a National Public Radio broadcast that reaches more than four million people weekly. I had just explained how evolution—the idea that mankind emerged by accident out of the primeval slime—leads to a view of human life that makes mass murder possible.

It is really not an exceptional claim. After all, if a loving, almighty God made man, then human life is sacred. But if evolution is true, then we are simply the product of time and chance, and there is no morality and no intrinsic worth to human life.

Evolution "liberates" man to do as he wishes—and to whom he chooses. Charles Darwin understood this. He wrote in his *Autobiography* that one who rejects God "can have for his rule of life, as far as I can see, only to follow those impulses and instincts which are the strongest or which seem to him the best ones."

And while Terry Gross may have her doubts, the two

most notorious and blood-soaked political movements of the twentieth century, Nazism and Communism, both rejected God and were animated by the idea of evolution.

It was Darwin's theory—carried to its logical conclusion—that led to the death of some 11 million people at the hands of German Nazis. Hitler was a devout evolutionist. He instructed his troops in evolution and had them provided with books by Darwin and Friedrich Nietzsche.

Hitler was determined to create a super race by eliminating the so-called inferior races. Following Darwin, he believed the Aryan was superior to other "races" (including the Jews, the Gypsies, the Slavs, and so on). Because of this, the Nazis made it illegal for Jews to marry other Germans. Hitler wrote in *Mein Kampf*:

> The stronger must dominate and not mate with the weaker, which would signify the sacrifice of its own higher nature. Only the born weakling can look upon this principle as cruel, and if he does so, it is merely because he is of a feebler nature and narrower mind; for if such a law did not direct the process of evolution[,] then the higher development of organic life would not be conceivable at all.[1]

Hitler tried to speed up evolution—to help it along.

"The German Führer, as I have consistently maintained, is an evolutionist," British evolutionist Sir Arthur Keith wrote in the 1940s. "He has consciously sought to make the practice of Germany conform to the theory of evolution."[2] And millions suffered and died in unspeakable manners because of it. Keith also said in his book, *Evolution and Ethics*, that "The leader of Germany is an evolutionist not only in theory, but, as millions know to their cost, in the rigor of its practice."[3]

Karl Marx wrote his *Communist Manifesto* before Darwin published his *On The Origin of Species*, but communism is nonetheless indebted to evolution. Marx, the founder of communism, found in evolution exactly what he needed: a pseudo-scientific foundation for his godless worldview. Marx wrote Friedrich Engels that Darwin's *Origin* "is the book which contains the basis in natural history for our view."[4] Marx also wanted to dedicate *Das Kapital* to Darwin but decided otherwise after Mrs. Darwin made known her displeasure at the association of her husband with Karl Marx.

"Darwin put an end to the belief," said Vladimir Lenin, "that the animal and vegetable species bear no relation to one another, except by chance, and that they were created by God, and hence immutable."[5] Communist henchman Josef Stalin became an atheist as a young man, while reading Darwin in seminary.

These communist leaders and others killed more people than all those killed in all religious wars combined. The "rough estimate" of the *Black Book of Communism* is that the "total approaches 100 million people killed."[6] Stalin, Mao, Pol Pot, and all the rest are the greatest mass murderers of all time—and all compliments of evolution.

Yes, ideas have consequences. And, in the case of evolution, much "fatal fruit."

But evolution is not just an idea that leads to bad consequences. It is also a false idea. Evolution is not a fact. It is not even a scientific theory. As the British astronomer, the late Sir Fred Hoyle, said:

> The notion that not only the biopolymers but the operating programme of a living cell could be arrived at by chance in a primordial organic soup here on the Earth is evidently nonsense of a high order.[7]

Tom DeRosa, Director of the Creation Studies Institute, has done a superb job of tracing the intellectual course by which Darwin's notebook jottings became fodder for Hitler's guns. He offers many examples throughout as to why evolution is false as well as toxic in its effect. It is a gripping and instructive account that deserves careful reading and wide attention. *Evolution's Fatal Fruit* is an eye-opening and sobering demonstration of why we, as Christians, must answer and refute the false claims of evolution.

It is my prayer that this book will be used to give one more push to the now crumbling and about-to-collapse edifice of evolution—what French biologist Jean Rostand has called a "fairy tale for adults."[8] In its place, may the wondrous effects of Christ's ethical, moral, and spiritual teaching once more prevail in our nation and in the world.

D. James Kennedy, Ph.D.

CHAPTER ONE

The Worship of Idols

Memories from youth washed over me as I listened to the rhythmic metal-on-metal clang of railcar wheels rolling through the endless dark subway tunnel. The New York City subway system is a complex network that connects five boroughs and provides millions of New Yorkers and visitors with transportation to work, school, entertainment, and sightseeing destinations. Years ago I lived and breathed the busy city air and rode the subway back and forth to college and work. Like so many city dwellers, it was my sole means of transportation.

The hard plastic seat and the steady, gentle, sleep-inducing rocking of the moving car felt very familiar. It was here in New York City that I was born and raised by caring Italian parents. I left the city in the late 1960s at age 21, but I was now back on assignment to tour the Darwin Exhibition at the American Museum of Natural History, located in upper Manhattan. The museum is, as it claims, "one of the world's preeminent scientific, educational, and cultural institutions."[1]

Charles Darwin's legacy is a topic of long-standing interest to me. His evolution manifesto, *On the Origin of Species*, was first published in 1859 and is said to be the most influential book in the world, besides the Bible. More than any other document, *Origin* has had the greatest secularizing influence in Europe and America over the last 150 years. It certainly had that effect in my own life as well.

I read *Origin* shortly after I completed my high school studies at a Roman Catholic seminary. I had been on my way to entering the priesthood, but *Origin* made me question my faith. It raised doubts that led me to become an atheist and denounce God while attending college in New York City. My return to the city and my long subway ride brought to mind how desperately I once wanted to be "free" in thought and action and how Darwin offered me an escape from God—one I eagerly took.

The Darwin Exhibition opened in November 2005,

and commenced a three-year celebration to honor the 200th anniversary of Darwin's birth and the 150th publication of *On the Origin of Species*. The exhibition is slated to travel next to Boston, Toronto, and Chicago and will reach the Natural History Museum in London in time for Darwin's 200th birthday celebration in 2009.

Visitors to the American Museum of Natural History in New York are greeted with a large sign announcing the special Darwin exhibit. The green and black sign features a photo of Darwin in old age. He has a sad and pensive appearance, with his long gray beard, aged wrinkled face, black hat, and dark coat. "Discover the man and theory that changed the world," the sign announces—a strong affirmation of Darwin's culture-wide influence in every public museum and public school in America.

The Darwin exhibition brings together many original manuscripts, including letters and notebooks that Darwin wrote on his famous five-year voyage on the H.M.S. *Beagle* and throughout his life in England. It also displays personal effects and artifacts that Darwin used, including those used to do his work as a naturalist on the *Beagle* from 1831-36 and until his death in 1882.

Niles Eldredge is curator of the American Museum's division of paleontology and curator of the Darwin exhibition. Eldredge, whose "main professional passion is evolution," according to his biography, wrote a paean to "St. Charles," *Darwin: Discovering the Tree of Life*, to accompany the exhibition and to celebrate Darwin's life and theory.

Both the exhibition and the book serve to educate the public on modern evolutionary biology. For Eldredge, who has battled creationism in lectures and books, the exhibit is an opportunity to combat what he calls "the rising tide of willful ignorance." The touring exhibition will, Eldredge hopes, serve to bring enlightenment to the evolution/creation controversy.

There should be no question from any serious biologist that life evolved, Eldredge reports. In the introduction to *Darwin*, Eldredge states that the "ultimate aim of this book and exhibition with which it is paired is the clear demonstration of the evidence and mode of thinking that led Charles Darwin to the conclusion that life has evolved."[2]

The Darwin exhibition is housed on the third floor of the American Museum's stately five-story landmark building, itself a nineteenth century heirloom. The American Museum was founded in 1869, and in 1877, it took up residence at its current location across the street from New York's famous Central Park.

The American Museum, known for its sponsorship of explorations on every continent on the earth between 1880 and 1930, excels in anthropological research. Its expedition to the North Pacific from 1897 to 1902 produced records that went beyond all expectations and gave new insights into the life and culture of the peoples living there. In 1926, the American Museum attracted a bright young student from Columbia University, Margaret Mead, who became a world-renowned anthropologist. She spent her entire life, until her death in 1978, at the Museum's department of anthropology.

Since the early 1920s, the American Museum has sponsored excavation and research of dinosaurs and other paleontological specimens in Mongolia's Gobi Desert, one of the best fossil sites in the world. Since 1991 there has been a joint venture with the Mongolian Academy of Sciences to continue these paleontological expeditions to the Gobi. These joint expeditions occur annually, adding more fossils to the Museum collections. Over the years the Museum has collected over a million vertebrate fossils, making its collection the largest in the world. Some 85 percent of the real fossils are in full view for all to see.

Evolutionary Propaganda on the Fourth Floor

Upon entering the Museum, I asked the elevator operator where to start my tour, and he quickly responded, "The fourth floor. That's where everybody goes. That's where the dinosaurs are." When I reached the fourth floor, all I could hear and see were excited children, accompanied by adults, as they went from one exhibit to another. Straight ahead was an archway at least two stories high leading into a long hall that had a large *Apatosaurus* dinosaur on the right side. It was surrounded by a staircase and platform you could use to walk around this large creature and get a sense of the tremendous size of these extinct beings. They could reach a height of 20 feet at the hips and had a body length of 70 feet from head to tail. The display also had an impressive set of large footprints in stone behind the creature's feet, which further indicate its immensity.

The fourth floor had two enormous halls filled with a wealth of fossilized material representing a wonderful variety of dinosaurs. The famous *T. rex*, every kid's favorite (mine, too), was prominently standing about 12 feet tall at its shoulder. It had a body length of 40 feet, stretching from its stiff tail at one end to its terrifying open mouth filled with long, sharp teeth at the other.

Just a few feet away, a *T. rex* skull was displayed at eye level in a thick, clear, plastic case, giving museum-goers a chance to stand face-to-face with the notorious, bony, five-foot head and its 50 serrated teeth. There seemed to be some sort of magnetic attraction to that *T. rex* skull because it drew many curious onlookers of all ages.

Things have really changed since I visited the museum as a young boy in the 1950s. The displays are now more engaging and allow much closer inspection. I certainly don't remember being able to go nose to nose with the dinosaurs in the past. No wonder the noise level is high and the enthusiasm in those voices is so contagious. The museum has succeeded in bringing families together and getting them excited about these old bones. That is a concern, since the displays are presented in a manner that makes them a persuasive and unremitting advertisement for the claimed "truth" of evolution.

Everywhere I looked in the dinosaur display area, there was evolutionary propaganda. The displays pointed out how the features of the different dinosaurs were similar, a demonstration of Darwin's idea that all living things, including dinosaurs, come from a common ancestor.

The dinosaur exhibits were designed to promote evolution through extensive renovations done to the museum during 1994-1996. The story of evolution takes center stage in the exhibit. Rather than show the dinosaurs in chronological order, the specimens are displayed "according to evolutionary relationships, dramatically illustrating the complex branches of the tree of life, in which animals are grouped according to their shared physical characteristics."[3]

In his 1993 blockbuster film, *Jurassic Park*, Steven Spielberg portrayed dinosaurs so realistically that audiences were kept at the edge of their seats. The raptors seemed to jump with fierce swiftness right off the movie screen. This film, and others that followed, produced a swell of media interest in dinosaurs. *Time* magazine gave readers "The truth about dinosaurs" in its April 26, 1993, cover story announcing, "Surprise: Just about everything you believe is wrong."

Time reported a plethora of new discoveries made in the 1980s that dramatically altered the state of knowledge about dinosaurs and evolutionary development. The implication was clear: Dinosaur paleontology is continually changing. It also raises questions: Will future discoveries give us better insights about these creatures? Should we make claims about the descent of dinosaurs from a common ancestor when all the evidence is not in? Will *Time* or another magazine tell us once more in the near future that when it comes to dinosaurs, "everything you believe is wrong"? This is something to keep in mind when one examines the claims of evolutionary paleontologists.

Most major museums, such as the American Museum of Natural History, are thoroughgoing in their belief in evolution, which is made abundantly clear in their exhibits. But is this stance based on scientific data, or is it derived from beliefs that precede the evidence and which predispose museum curators toward evolution? One wonders, since an honest look at the fossil record shows that it is grossly incomplete.

It is not known, for example, how many kinds of dinosaurs once existed. *Discovering Dinosaurs*, a book published by the American Museum of Natural History in 1995, attempts to answer this question with a nebulous estimate. Based on a "literal reading of the record" during the late Cretaceous period (said to be about 65 million years ago), the last time dinosaurs roamed the earth, there were only some 40 different species.

"Because the fossil record is incomplete, we can view the record only as a minimum estimate of the number of kinds of dinosaurs that existed in the past," write the authors.[4] The real question is: How can you be sure, if you are looking at an incomplete fossil record? Are there just a few fossils missing, or is there a myriad number? The authors admit: "We do not know how bad the fossil record is, but we suspect it may be very bad."[5]

If the fossil record is suspect, how can we form conclusions and adorn museums with evolutionary diagrams that claim to be scientific truth? The only answer is a well-established belief system based on Darwin, who is lionized on the third floor of the American Museum of Natural History.

"It Will Change Your Life"

The foyer outside the Darwin exhibition is so dimly lit that one's eyes are inevitably focused on the large, brilliantly illuminated glass cases in which creatures and plants from the Galapagos Islands are displayed. These plants and animals are typical of what Darwin must have observed when he set foot on Galapagos during the *Beagle* voyage. Live Galapagos tortoises lumber about inside one case. As visitors wait, they get a sense of what Darwin saw and what influenced him to formulate natural selection as the mechanism for evolution. It was on Galapagos that Darwin observed incredible diversity in the flora and fauna, as he went from one island to another. One gets the sense that we are standing in Darwin's shoes, looking over the bow of the *Beagle*, as we patiently wait for the line to move forward.

At the exhibition entrance, a gigantic picture of Darwin, clad in brown coat and sporting his full white beard, greets visitors. DARWIN, in bold gold letters, is splashed across the portrait, under which visitors are given a hint of what is to come:

> Happiest at home with his notebooks and his microscope, he shunned the public eye. Controversy made him ill. This brilliant observer of nature kept his most original and revolutionary idea under wraps for decades. Yet today, two centuries after Charles Darwin's birth, nearly everyone knows his name. What did Darwin do, and why does he still matter so much?[6]

Duly instilled with a sense of mystery and awe, visitors tread into the exhibit, which fills much of the capacious third floor. A reverential mood spreads over the crowd. The faces of many radiate a solemnity often associated with a place of worship. This exhibit is, for many, a spiritual journey to examine a man whose ideas have been idolized for 150 years. This was made obvious when I approached the docent and asked her how much time it would take to tour the whole exhibit. In soft, but unmistakably passionate tones, she said, "If you read everything, an easy 90 minutes, and it will change your life." Her answer made it obvious how Darwin influenced her life and how important evolution is to her. I believe that most of the people with me on the tour felt the same way.

One of the first exhibits shows a large magnifying glass and an aged reddish-brown notebook. Nearby text tells visitors that, "Charles Darwin looked closely at life." Indeed, Darwin was a collector of a variety of plants and many invertebrates and took copious notes throughout his life. It was appropriate to begin with the tools that Darwin used as a naturalist, but these were mere props for the exhibit's true pedagogical intent. Again, the adjacent text offered a subtle clue of what was to come. We learned that, for Darwin, the "vast and marvelous diversity of life on Earth, from barnacles to butterflies, ostriches to orchids, made him curious."

The Question Darwin Couldn't Answer

Those poetic words are followed by three suggestive questions that easily lead to answers that favor Darwinian evolution.

1. Countless species, living and extinct—
   why so many?
2. Some were only slightly different from
   one another—what could explain that?
3. Each organism so beautifully adapted to
   its environment—how could it happen?

I wonder if the people who pay homage to Darwin at the museum recognize that Darwin could not define "species," the term on which his revolutionary theory was based. The exhibition refers to "countless species, living and extinct," but neither Darwin nor contemporary biologists have been able to reach a commonly held definition of this term. How then can we know when one living thing is a member of the same, or a different, species?

It might be better to refer to the "countless variety" of living forms. But then, what about those "species" which are "only slightly different from one another"? Aren't people, in fact, "slightly different" from each other? Does that mean then, that redheads are a different species from brunettes and blondes? Do men over 250 pounds constitute a different species than the ectomorphs among us—

guys who belong to the bantam weight class? What is meant by "slightly different," and when are slight differences enough to constitute a separate species?

The exhibition questions listed above are premised on the evolutionary idea that slight differences will, over time, add up to big differences—making it possible for microbes to, in time, evolve into men. These questions presuppose the absence or non-existence of the Creator, the One who perfectly designed all living forms to live in their environments. They also lead to the conclusion that every creature has a common ancestor and that differences between living forms come not from a Creator God but from "Natural Selection."

Do exhibition visitors understand that these questions about the number, variety, and adaptation of living things are not answered by Darwin or his present-day evolutionary followers? While he used the term in his title, Darwin never defined "species" in his book, *On the Origin of Species by Means of Natural Selection*, first published in 1859. The fundamental question Darwin sought to answer was: Where did the species come from? But he never defined that most important term—species.

"Every Naturalist Knows Vaguely"

Not only that, Darwin actually avoided defining the term because of widespread disagreement over its meaning. He stated in all six editions of *Origin*:

> Nor shall I here discuss the various definitions which have been given of the term species. No one definition has as yet satisfied all naturalists; yet every naturalist knows vaguely what he means when he speaks of a species.[7]

In many of the latter editions of *Origin*, Darwin provided a glossary of terms in which the term "species" is noticeably missing. Darwin circumvented the definitional problem by stating that, on this matter, "the opinion of naturalists having sound judgement and wide experience seems the only guide to follow."[8]

Not only was the word species undefined by Darwin in 1859, it remains undefined today, some 150 years later. Educators do give students an oversimplified and unrealistic definition for species, but when you check the scientific technical literature, there is no consensus on this key term:

> What are biological species? At first glance, this seems like an easy question to answer. *Homo sapiens* is a species, and so is *Canis familiaris* (dog). Many species can be easily distinguished. When we turn to the technical literature on species, the nature of species becomes much less clear. Biologists offer a dozen definitions of the term "species" (Claridge, Dawah, and Wilson

> 1997). These definitions are "not fringe accounts of species but prominent definitions in the current biological literature."[9]

Evolution is defined as a dynamic force that by means of natural selection will change living forms by tiny steps over long periods of time. The National Academy of Science defines evolution as "Change in the hereditary characteristics of groups of organisms over the course of generations."[10] This assumption of continual change through many generations makes it difficult to establish boundaries in defining a species. Paleontologists trace life's history by using fossils of animals and plants. They often disagree and revise their classification of fossil specimens. For example, there are different species of mammoths. The California dwarfs (midget mammoths at 4-6 feet), Imperial (very large at 17 feet tall), Columbian (larger than an elephant at 14 feet), and Wooly (with long hair and the size of smaller elephants) run into each other on the evolutionary time line as they go from one generation to another. There is much discussion as to which one goes where on the evolutionary time line. This is an example of the difficulty encountered by paleontologists in deciding the boundary of a species. There seems to be never-ending deliberation and debate as to where a species belongs. It is always changing.

This is the problem Darwin has passed on to

biology. What constitutes a new species if the species is in constant flux? If the physical characteristics that were once labeled a species disappear and then return after many generations, there is confusion in the evolutionary coding system. *The Stanford Encyclopedia of Philosophy* reports:

> Biologists have had a hard time finding biological traits that occur in all and only the members of a species.... All it takes is the disappearance of a trait in one member of a species to show that it is not essential. The universality of a biological trait in a species is fragile.[11]

What about the great variety of living creatures? Did this arise from natural selection, or did it come from information placed in the genetic code? There are currently over 400 known breeds of dogs.[12] These dogs come to us in varying heights, weights, shapes and forms. The differences can be extreme: the Chihuahua stands just a few inches above the ground, while the Irish wolfhound stands roughly three feet tall.

When we observe what is common and what is unique in organisms, there will always be questions. How different kinds of animals interbreed now is not necessarily how they bred in the past. No wonder the evolutionists disagree on a common definition for "species."

Precious Little Proof

One fact not presented to patrons of the Darwin Exhibition is that the man who is honored there had precious little proof in the fossil record to support his claims. Nonetheless, museums often attempt to prove evolution by showing varieties of the same kind. For instance, I witnessed a wonderful display of trilobites—an extinct, bottom-dwelling ocean creature—of all different sizes and shapes. But what I observed was a variety of trilobites, not evolution. For evolution to be proven in the fossil record there have to be fossils that mark the transition from one kind to another.

Darwin knew there were gaping holes in the fossil record, but hoped that in time, new discoveries would be made to fill in the gaps and establish the evolution continuum. It was Darwin's belief that slight, minute changes over a long period of time would turn bacteria into *homo sapiens*. The fossil record offers no evidence to support that assertion. The late Harvard paleontologist and evolutionary biologist, Stephen Jay Gould, a well-known defender of evolution, admits that:

> The extreme rarity of transitional forms in the fossil record persists as the trade secret of paleontology. The evolutionary trees that adorn our textbooks have data only at the tips and nodes of

their branches; the rest is inference, however reasonable, not the evidence of fossils.[13]

For Darwinian evolution to be proven unequivocally true, the fossils should produce a record that would give an unbroken testimony. They do not. If one observes the fossils buried in the rocks, there is no evidence of simple life forms evolved into more complex forms. Actually what we see is just the opposite. Dr. John Morris, President of the Institute for Creation Research, notes that, "The fossils at the bottom (i.e., long ago) are equally as complex as any animal today and are essentially the same as their modern counterparts."[14] What you really see is fully formed fossils abruptly buried in strata without any of their "less adapted ancestors in lower levels that would have preceded them in time."[15]

After Darwin's voyage on the H.M.S. *Beagle*, he asked Richard Owen in 1837 to identify some of the specimens he collected on the Galapagos Islands. Owen was a leading British anatomist who believed in an outside intelligence, a Creator of the universe. A taxonomist, he is famed for giving the name "Dinosauria" to what he identified in 1842 as "a distinct tribe or sub-order of Saurian Reptiles."[16]

Owen introduced the term "homology," which he described as "a common structural plan for all vertebrates." These common structures were called "archetypes." He believed these archetypes were designed by "the Divine mind," who "foreknew all its modifications." [17] He stated very clearly that each species had its own "organizing energy" that would dictate how far the species would change within strict limits. Owen did not believe in evolution from one kind to another and was emphatic with young Darwin about this. He pointed to "the Divine mind" as the organizing force, warned Darwin against proposing the "transmutation of species" (another term for evolution), and became one of Darwin's strongest antagonists when Darwin went public with *Origin* in 1859.

Is homology—the presence in different kinds of common structures, such as skeletons, organs, and vascular systems—a proof for evolution? If so, one might also conclude that the wheel, a common element in bicycles, motorcycles, cars, and eighteen-wheelers, is also evidence of evolution. In point of fact, however, the wheel is the product of design. The wheels in the instances above have been modified to meet slightly different demands—something called for by their designers. Any good designer will repeatedly utilize simple structures to accomplish his design purpose. So, homology is not evidence for evolution, but for design.

The Darwin exhibition is premised on the assumption that there are scientific evidences to prove evolution. As the exhibition tells it, it is the genius of the humble and quiet Charles Darwin that saved the world from ignorance. He is celebrated because he discovered the means by which all living things came to be—the eons-long trail of death known as natural selection. The message is loud and clear: Charles Darwin changed the world because he found a way to do away with a Creator God.

Darwin's theory effectively dismisses the Creator God, who made each according to its kind (Genesis 1:24). God is not needed and has been replaced. As I looked around at the faces of exhibition visitors, many who had come long distances to this spectacular presentation, I wondered how many know or have the open-mindedness to consider the myriad problems with Darwin's theory of evolution.

The three questions above that introduced this exhibition cannot be answered by what Darwin proposed. I came to the exhibit prepared, knowing the massive problems that beset the theory of evolution and was curious to see if exhibit organizers would be honest and open about the growing chorus of objections to Darwin. They were not. Those three questions have never been answered by evolution. Richard Owen was correct in his conclusion that the evidence points to a Divine Mind, an Intelligent Creator.

Dismissing God and the Bible

As I moved about the exhibition I was fascinated with all the different manuscripts that were penned by Darwin.

In the middle of the exhibition, an impressive display replicated Darwin's Down House study in England. It was here, from 1842 to 1882, that Darwin developed his research, conducted experiments, and wrote many of his books, including many subsequent editions of *Origin*.

Most artifacts, like his dissection utensils, magnifying glass, microscope, and other scientific paraphernalia were arrayed on a wooden desk in an orderly manner. His study gave the appearance of industry and organization—a reflection of Darwin's personality. I could imagine Darwin sitting at that desk for long hours writing and thoroughly engaged without distraction. He was a goal-oriented man of habit, who loved working in his study, collecting specimens, or taking patient care of his plants. He basically never gave up his life's work as a naturalist that began when he was a young man.

The exhibition seemed to me to capture the ambiance of how Darwin lived, with its old manuscripts and dog-eared notebook covers. But I also felt uncomfortable, because the exhibition also served to dismiss God and the Bible.

The first section of the exhibition looks at Britain's social milieu, "Before Darwin." Before *Origin*, people were said to believe that 1) species were not linked by common descent in a family tree of life; 2) species never change; and 3) the earth is just 6,000 years old, so "there would not have been time for species to change." It is interesting to note that these pre-Darwin premises all employ the scientific term species which, of course, cannot be acceptably defined. One wonders why it is used so often in the exhibition and, in fact, why it is used at all.

The exhibition tells visitors that the people of England, with their simple faith in the biblical creation account, lived in a state of naïve ignorance. They were not asking intelligent, probing questions like:

> How could a new species arise? Few naturalists, however, were even asking such questions. Most were comfortable with the prevailing view that each species resulted from an act of special creation by a Creator. Why didn't more people grasp that similarities in skeletal structures—so clearly visible—were a clue that species are related?[18]

I was distressed at this public attack on Christianity and the Bible on the third floor of the American Museum of Natural History, coupled as it was with a wholly uncritical treatment of Darwin and his theory. The exhibition made, for example, this bold assertion about Christians blinded by their faith in Genesis:

> In England during the 1700s and early 1800s, few questioned the Biblical story of creation. The

> prevailing view was that people were created to rule over animals, "over the fish of the sea and over the birds of the sky." Naturalists could see that, particularly beneath the skin, a chimpanzee looks a lot like a human—but the idea that we might somehow be related to apes was almost unthinkable.[19]

The takeaway message from this remarkable Darwin display was what I had expected. Yes, it is a celebration of Darwin and his achievements, but more than that, it tells visitors that mankind's ancient enmity against God is made manifest in the enthusiasm for Darwin and his theory.

Although Darwin's theory of evolution is presented as science, it is not. After all, if "species" is left undefined, how can you intelligently discuss its origin and subsequent diversification, and call that science? In fact, evolutionary science is in a shambles.

What I witnessed on the third floor of the American Museum was a masterful charade—the idolization of Darwin and his theory. The huzzahs for Darwin, I believe, were generated not so much by his ideas, but for their consequences—to liberate man from God. This is eloquently said by the curator, Niles Eldredge. In his companion book, *Darwin: Discovering the Tree of Life*, he describes Darwin as the leading agent of secularization in

Western society:

> Darwin split the scientific world off from the religious world, at least as far as the "mysteries of mysteries" was concerned. He convinced the scientific world—and beyond that, the thinking world—that life has evolved through natural causes. Because he was so careful in building the case for evolution, and for evolution of human beings, Darwin did more to secularize the Western world than any other single thinker in history.[20]

I loved visiting the place where I was born and raised. New York City has so many great things to see, like the American Museum. Riding the subway turned my mind to many pleasant memories, but the message delivered at the Darwin exhibition left me very unsettled. It was a frontal attack on the personal faith I have in Jesus Christ—an attack based not on scientific evidence, but on sinful mankind's long-standing hostility toward his Creator.

CHAPTER TWO

Nurturing Charles Darwin

Charles Darwin replaced the benevolent Creator with natural selection as the means by which all life came into being. His elaborate theory was a magnet to those who were eager and willing to displace God with a scientific construct that left man free to act as he wishes. Evolution was, in fact, a tool to do away with God. After all, if science tells us that everything living came into being by means of natural selection, what need is there to pay heed to the claims of the God of Scripture?

Darwin published *On the Origin of Species* at a time when Victorian England was a predominantly Christian nation. Generally, the British were serious about their faith, but as Darwin's theory gained acceptance, faith declined. It is estimated that only 20 percent of the British population are active Christians at present.[1]

Darwin's appeal remains undimmed nearly 150 years after he published *Origin*. His devout admirers continue to shower accolades on him for providing what is to them a plausible replacement for the Creator God. It is no wonder that Darwin's name and picture adorn almost every high school biology textbook. For example, Darwin defender Kenneth Miller, coauthor of *Prentice Hall Biology*, gives his young readers a chapter-length treatment on Darwin's personal observations and beliefs. Evolution is taught in every state in the union and throughout the world as a fact of science—one no less so than gravity or atomic structure.

Many of Darwin's devotees share the passion of Niles Eldredge, who has exclaimed that, "No idea in science has shaken society nearly so much as evolution—and in making that claim. I include everything, even $e= mc^2$."[2]

So who is this man? What motivated him to turn the world upside down with his bold proclamation that humanity descended from the "beasts of the field?" Why did he cast away his original belief in a loving Creator responsible for order and design in the universe for the dreadful, impersonal, and purposeless state of affairs described in his theory? Darwin answered the who, the what, and the why questions in his voluminous paper trail of personal correspondence, notebooks, published books, and scientific papers.

Charles seems to have always had a pen in his hand. A copious note taker, he had an eye for detail, as a trained naturalist would be expected to have. Darwin would not think twice about writing notes on the margins of the books he studied. He would even rip apart the binding of large, cumbersome books, so he could carry them around with him. More than 15,000 personal manuscripts of Darwin's are housed at Cambridge University.

Francis Darwin, the seventh of Darwin's ten children, published his father's letters in 1887, with a second edition in 1903. Darwin's *Autobiography*, which he finished in 1876, was intended just for his family, but Francis and the family decided otherwise and published it in 1892. Darwin's detailed and well-organized correspondence and autobiography are easy-to-follow guides to this man's life and thought.

Charles Darwin was born on February 12, 1809, in Shrewsbury, England, into prestige and wealth. He was raised in a very large house where his father, Robert, an extremely successful physician, treated his patients. Darwin's father enjoyed the respect and love of the community and reportedly had the highest income of any provincial doctor.[3]

Charles remembered his father with deep fondness, writing of "my father, who is the kindest man I ever knew, and whose memory I love."[4] He observed in his father a man who knew medicine on a practical level, had common sense experience, and was sensitive to the needs of people. This was a formula for a successful medical practice that brought in a great deal of revenue. An investor, Robert Darwin also had a "remarkable head for business."[5]

Darwin's grandfather, Erasmus, Robert's father, was a gifted intellectual and talented physician, as well as a poet, philosopher, botanist, naturalist, and inventor. He had the intellectual capacity to master almost anything to which he set his mind. He was enamored with the creative engineering genius of inventors and scientists of his time. He rubbed shoulders with these intellects and wrote a book, *Darwin's Commonplace Book*, which contained imaginative ideas not then patented. Many of his ideas became reality, including copy machines, rocket motors, and a canal lift. Gifted and gargantuan—his corpulence made it necessary to refashion his dining room table to accommodate his girth[6]—Erasmus died in 1802, seven years before grandson Charles was born.

Charles' mother, Susannah Wedgwood, was the daughter of Josiah Wedgwood I, the great industrialist, inventor, and founder of the famous Wedgwood ceramic ware firm. The Wedgwoods and the Darwins have a long family history in business and marriage. Erasmus Darwin and Josiah Wedgwood I were close friends. In 1796, Erasmus' son, Robert, married the oldest Wedgwood daughter, Susannah. Robert and Susannah had five children, with Charles being next to the youngest. Darwin's mother died at the young age of 32, when Darwin was only eight years old. He had little memory of his mother, and in his latter life said, "It is odd that I can remember hardly anything about her except her death-bed, her black velvet gown, and her curiously constructed work-table."[7]

Susannah was often ill and bedridden and was not directly involved in the rearing of her son. Darwin enjoyed his childhood, even though he lost his mother at an early age, perhaps because his older siblings helped fill the maternal gap. He had a strong bond with his two older sisters and an especially strong bond with his older brother, Erasmus.

The Darwins and Religion

Susannah and her children attended a Unitarian church in Shrewsbury, where a memorial marble plaque notes of Charles Darwin, "In an early life, a member and a constant worshipper in this church."[8] In her faith, Susannah followed her father, Josiah Wedgwood I, who became a Unitarian in the late 1760s and exercised a strong influence on his family's religious beliefs.

Although there is no set doctrine in the Unitarian

church, and each church is encouraged to be unique in worship and style, much of orthodox Christian belief is dropped. Unitarian belief was focused on the right and potential of individuals to improve themselves and others without reliance on the Bible. Unitarianism rejects the deity of Christ, the Trinity, and the fall of man. It holds that man is basically good and should not be judged to Hell. It was considered a non-conformist body that dissented from the established Church of England. Even though Charles attended a Unitarian church, his father had Charles baptized in the Church of England.

Charles Darwin married his first cousin, Emma Wedgwood, who faithfully followed her family's strong Unitarian tradition. This unique marital bond gave both Charles and Emma the same Unitarian grandfather, Josiah Wedgwood I. Charles continued to attend Unitarian services with his sisters on a regular basis after his mother's death and then attended regularly with his wife during the first ten years of their marriage. But after the death of his dearly loved daughter, Annie, his second child, at age ten, he swore off worship and God. Annie, Darwin's favorite, succumbed to the painful illness of tuberculosis in 1851. After that, Darwin never came to church for worship again. He gave up on the idea of eternity and Heaven, writing about his daughter's death that, "Our only consolation is that she passed a short, though joyous Life."[9]

Darwin's education began in 1817 at age eight in a

grammar school where his local Unitarian minister served as the teacher. Later, Charles joined his older brother Erasmus at the Rev. Samuel Butler's prestigious boarding school. The classical curriculum taught there required Darwin to learn Greek and other languages, which he struggled mightily to master. "I could never do well," he admitted. "I have been singularly incapable of mastering any language."[10] A great deal of memorization was required, which Darwin said he would easily forget within the next 48 hours. For him, Rev. Butler's school was a waste of time. "Nothing," he wrote, "could have been worse for the development of my mind."[11]

Charles' father took exception to his son's academic performance. He told his son, "You care for nothing but shooting, dogs, and rat-catching, and you will be a disgrace to yourself and all your family."[12] A budding naturalist, Charles was bored at school, but loved to hike and collect minerals, shells, insects, rocks, and plants. His interest in nature was fueled by books, such as *Wonders of the World*, which he believed motivated him to later venture onto the H.M.S. *Beagle* for a global tour.

The "Lunatics" Society

Both of Charles' grandfathers, Erasmus Darwin and Josiah Wedgwood I, held membership in a social club of intellectuals, scientists, and industrialists who met monthly on the Monday closest to the full moon (to ensure adequate light to make their way home at night). The Birmingham Lunar Society, a think tank for the Industrial Revolution, had as its goal the improvement of mankind on a materialistic and philosophical basis.

These eighteenth century "movers and shakers" constituted a gathering of influential elites. British physicist Dr. Desmond King-Hele has said that the members of the Lunar Society "are now seen as leading British contributors to the Enlightenment, so much more impressive than the French philosophers because they were practical instead of being merely intellectuals preaching from ivory towers."[13] Other members included James Watt, the inventor of the steam engine; Matthew Boulton, steam engine manufacturer; and Joseph Priestley, a Unitarian preacher, philosopher, and amateur chemist, who discovered carbon dioxide, carbonated water, oxygen, and laughing gas. Benjamin Franklin also made frequent visits to the meetings of this group.

The Lunar Society supported attempts to free man from government tyranny and absolute law. They agreed with the aims of the French and American revolutions. Many of these "lunatics," as they referred to themselves, denied God. Erasmus Darwin called the deity the undefined first cause and Joseph Priestley was the most important proponent of Unitarianism in eighteenth century England.

LIKE GRANDFATHER, LIKE GRANDSON

Erasmus Darwin, the prime mover of the Lunar Society, was a confirmed evolutionist. He believed the following:

> . . .that all warm-blooded animals have arisen from one living filament, which the great First Cause endued with animality, with the power of acquiring new parts, attended with new propensities, directed by irritations, sensations, volitions and associations, and thus possessing the faculty of continuing to improve by its own inherent activity, and of delivering down these improvements by generation to its posterity, world without end.[14]

Darwin's grandfather had a theory of evolution that was consistent with what his grandson proposed 65 years later and what modern evolutionists believe today. However, the supposition that biological change is upward has never been observed in the laboratory or in the field. To add limbs to an organism, such as a fish, and make him a warm-blooded cow is nothing but fantasy and absurd.

Erasmus Darwin, a great intellect of his times, certainly follows this path of origins foolishness in his *Zoonomia; or The Laws of Organic Life*, published in two exhaustive volumes in 1794 and 1796. These practical guides for physicians described and classified prevalent diseases of the time and prescribed treatments. This major work was very much in demand and went through a number of editions and translations and even made its way across the Atlantic to America. Erasmus included his evolutionary suppositions throughout the publication, which became the first comprehensive expression of the evolution hypothesis. It also gave vent to the elder Darwin's caustic antagonism toward Christianity. He included in his catalogue of diseases, "Credulity, Superstitious Hope, and the Fear of Hell."[15]

Erasmus Darwin's 1802 poem, "The Temple of Nature", demonstrates his evolutionary worldview and sounds a theme that organic life came from the inert ocean, an idea to which his grandson later turned.

The Temple of Nature
Organic Life beneath the shoreless waves
Was born and nursed to Ocean's pearly caves;
First forms minute, unseen by spheric glass,
Move the mud, or pierce the watery mass;
These, as successive generations bloom,
New powers acquire, and larger limbs assume;
Whence countless groups of vegetation spring,
And breathing realms of fin, and feet,
and wing.[16]

Erasmus Darwin never explained how one species evolved into another but, like his grandson, referred to the competitive struggle for life and sexual selection. Charles Darwin is hailed as a hero for identifying natural selection as the mechanism for evolution, but it appears that his grandfather was saying much the same decades earlier, albeit without the catch phrase, natural selection.

While Darwin read his grandfather's *Zoonomia,* he claims it failed at "producing any effect on me." At the same time, he acknowledges that "hearing rather early in life such views maintained and praised may have favoured my upholding them under a different form in my *Origin of Species.*" As to *Zoonomia* itself, Darwin said he was "much disappointed" on his second reading, a decade or so later—"the proportion of speculation being so large to the facts given."[17]

Darwin's critique of his grandfather has been widely echoed by critics of *On the Origin of Species.* William Jennings Bryan, an early twentieth century foe of evolution, eloquently said:

> Give science a fact and it is invincible. But no one can guess more wildly than a scientist, when he has no compass but his imagination, and no purpose but to get away from God. Darwin uses the phrase 'we may well suppose' 800 times and wins for himself a high place among the

> unconscious humorists by his efforts to explain things that are not true.[18]

Was Charles Darwin influenced by his grandfather? Creationist writer Russell Grigg notes that "almost every topic discussed and example given in *Zoonomia* reappears in Charles' *Origin*."[19] Indeed, when Charles began to formulate his theory of evolution in 1837, he wrote the title of his grandfather's book, *Zoonomia*, at the top of his notebook. One example of the debt Charles owes to Erasmus is found in his treatment of birds and their beaks. Grigg points out that:

> One of Charles' chief arguments for evolution is based on the shape of the beaks of finches in response to the types of food available that he saw in the Galapagos Islands in 1835. Is it credible to think that he had not been influenced by what Erasmus had written on the subject? Namely: "Some birds have acquired harder beaks to crack nuts, as the parrot. Others have acquired beaks adapted to break the harder seeds, as sparrows. Others for the softer seeds of flowers, or the buds of trees, as the finches.... All ... gradually produced during many generations by the perpetual endeavour of the creatures to supply the want of food."[20]

Both Erasmus and Charles were going down the same path. They were seeking to answer the question of where all things originated by means of evolution. Erasmus first made his views known in 1770 when he created a family coat of arms featuring three scallop shells and the phrase: "*E conchis omnia*" (everything from shells). Erasmus painted his evolutionary coat-of-arms on his carriage, where it was noticed and satirized by Canon Seward of Lichfield Cathedral. The Canon wrote that Darwin

> ... renounces his Creator
> And forms all sense from senseless matter.
> Great wizard he! by magic spells
> Can all things raise from cockle shells.[21]

Both Erasmus and Charles had the same thing in mind—a theory to remove the personal God from His creation.

Drifting Without Purpose

It was Darwin's father, Robert, who suggested that his son should go to medical school, after observing him as he helped in his medical practice. Charles wrote, "My father, who was by far the best judge of character whom I ever knew, declared that I should make a successful physician—meaning by this, one who gets many patients."[22] Dr. Darwin wanted Charles to follow his brother Erasmus to medical school at Edinburgh, Scotland, one of the most renowned medical schools in Europe.

Charles spent two years at Edinburgh University in medical school and found the lectures "intolerably dull." He also had a serious problem stomaching surgery, which distressed him a "good deal." He writes that he "saw two very bad operations, one on a child, but I rushed away before they were completed."[23] Of course the operations were done without anesthesia. One can only imagine the pain and piercing shouts that would fill the air. Darwin realized this was no place for him.

While at Edinburgh, Darwin found himself hiking around Wales and learned how to make observations and record his notes in a field notebook—skills essential to his tenure on the H.M.S. *Beagle*. He also became a member of the Plinian Society, a natural science club in which students presented research papers and hosted open discussions on the papers. The mainstay of the society was a young biology professor, Dr. Robert Grant. Darwin became close friends with Grant, known as a freethinker who was open and enthusiastic about new origins ideas. Darwin described Grant as a fellow he knew well "with much enthusiasm beneath this outer crust."[24]

Grant entitled his doctoral thesis, *Zoonomia*, in honor of Erasmus Darwin's evolutionary treatise. He was an enthusiastic advocate of Jean de Lamarck's theory of organic evolution through acquired characteristics. Darwin became a devoted student of Grant, assisting him in the collection of specimens and gleaning information and knowledge from his master.

Some years later the warm relationship between Darwin and Grant dissolved, after Grant was attacked for "blasphemous derision of the truths of Christianity." It may be that Darwin, having returned from his *Beagle* exploration, maintained a distance to keep his work untainted by the controversial Grant.[25]

After two years at Edinburgh Medical School, it became evident to Charles' father that his son was not going to become a doctor, so he suggested that Charles become a clergyman. Darwin had reservations, writing, "I had scruples about declaring my belief in all the dogmas of the Church of England."[26]

Darwin said he read, "with great care Pearson on the Creed and other books on divinity."[27] This was a not inconsiderable investigation, since British theologian John Pearson's *Exposition of the Creed*, a standard and celebrated work in the Church of England, ran, in its 1864 revised edition to 700 pages.

After his careful examination, Darwin was convinced. "As I did not then in the least doubt the strict and literal truth of every word in the Bible, I soon persuaded myself that our Creed must be fully accepted."[28] Darwin complied with his father's proposal, because he had a great deal of faith in his father's opinion. Dr. Robert Darwin was not a churchgoer and had very little concept of a clergyman's duties, but for reasons that are unclear, he saw his son as a man of the cloth. While Robert's thoughts are unknown, it is evident that his son took the proposal seriously.

Charles began his studies in 1828 at Christ Church College, part of Cambridge University, and took three years to complete his degree. Once again, he did not apply himself and only attended the required lectures. He felt his "time was wasted as far as the academical studies were concerned."[29] Charles seemed to do best when he was engaged in collecting, dissecting, touching, organizing, and writing, but not memorizing. He spent a great deal of time shooting fowl in the countryside, learning how to be a collector of beetles, and socializing with his friends.

His interests were not in the Scriptures or sharing the Gospel with others. Of all the coursework he took in those three years, Darwin mentions just two that he enjoyed and took pleasure studying: Euclidean Geometry and William Paley's *Natural Theology*. Darwin mastered these classes but barely passed his classical courses. Even so, he graduated tenth in his class, not bad for one who did not apply himself in three years at Cambridge.

Even with passing grades and earning his degree in 1831, Darwin never became the country parson his father proposed. He had a passion to study God's Creation, fueled by his reading of Paley's *Natural Theology*, a work which offered, as the subtitle stated, *Evidences of the Existence and Attributes of the Deity*. Darwin, whose thinking soon radically altered, was so enamored with the text that, "I could have written out the whole of the evidences with perfect correctness."[30] Several years later, however, after returning home on the H.M.S. *Beagle* in 1836, Darwin had rejected Paley and embraced his theory of evolution.

CHAPTER THREE

Darwin's Rebellion

Darwin's three-year academic slog through Cambridge might have ended early in failure, as had his time at Edinburgh, except for John Stevens Henslow. This young natural science professor offered Darwin the chance to do what he loved, natural science. Yes, there were still dreary lectures to attend and reams of what, for Darwin, was meaningless information to commit to memory. He had tried and failed to become a doctor, and now young Darwin was on track, albeit without passion, to become an Anglican clergyman. But his career path was altered once more—this time permanently—after Henslow took Charles under his wing and opened up to him an exciting new world that he enthusiastically embraced.

Henslow shared with Darwin his expertise, knowledge, and passion for hands-on specimen collection. He showed his young charge how to classify plants, insects, rocks—anything that could be observed outdoors in nature. It was not part of Darwin's Cambridge curriculum, but this extracurricular experience with his favorite mentor redirected the course of his life. Darwin wrote that Professor Henslow was the "circumstance which influenced my whole career more than any other."[1]

John Stevens Henslow came to Cambridge in 1822 and took the professorship in mineralogy as a young, energetic professor who loved to engage his students. Erasmus, the older brother, wrote to Charles from Cambridge, informing him that the Henslow lectures were entertaining. When Darwin arrived at Cambridge he sought out Henslow who, by 1828, had taken the Botany Chair at the University. Darwin praised Henslow for the clarity of his lectures, which contained "admirable illustrations."

Darwin also relished the opportunity to go on Henslow-led field trips. The professor, Darwin wrote, "used to take his pupils, including several of the older members of the University, on field excursions, on foot or in coaches to distant places, or in a barge down the river and lectured on the rarer plants and animals which were observed." Darwin loved to hike with Henslow. For Darwin, "these excursions were delightful."[2]

Henslow would also open his house every Friday night for informal discussions on various science topics for the undergraduates. Charles Darwin made his appearance and literally soaked up every morsel of knowledge Henslow threw out. Darwin saw in Henslow what he wanted to become and built a friendship that would last a lifetime.

Darwin's rapt absorption with his mentor was so pronounced that at Cambridge he became known as "The man who walks with Henslow." The professor was a true teacher, who made an investment in Charles and taught him all he knew about plants, rocks and rock layers, math, insects, chemistry, and much more. Henslow both inspired and passed on to Darwin his passion for natural science.

Henslow, who served as Darwin's tutor in his last year at Cambridge, was not just a naturalist, but a pastor with great skill in dealing with people. He was an ordained minister of the Church of England, who had received Holy Orders in 1824 and had a heart to serve God in ministry. He became a rector at Hitcham in Suffolk in 1837 and was a true shepherd for his sheep, taking care of his parishioners' spiritual and emotional needs.

Darwin saw in Henslow an honest and giving man who lived out his commitment to Christ. Darwin wrote of him:

> He was deeply religious, and so orthodox that he told me one day he should be grieved if a single word of the Thirty-nine Articles were altered. His moral qualities were in every way admirable. He was free from every tinge of vanity or other petty feeling; and I never saw a man who thought so little about himself or his own concerns. His temper was imperturbably good, with the most winning and courteous manners;...[3]

Although he did not leave an autobiography or copious correspondence, as was the case with Darwin, a published sermon by Henslow offers insight into this scholar-parson. He believed that the message of Christ should be available to all:

> I believe the "simplicity that is in Christ" to be as much the property of the unlettered as of the learned, the method of comprehending it being the work of the Spirit, and not of the understanding.[4]

Adam Sedgwick proved to be another major

academic influence on Darwin at Cambridge. Sedgwick became the Woodwardian Professor of Geology at Cambridge in 1818 and held that post for 55 years. He was a young charismatic geologist and a close friend with Henslow. The two were so closely matched that, "in religion, love of truth, and hatred of wrong, they were in exact agreement."[5]

Sedgwick was a dynamic and exciting geology professor who presented spellbinding lectures that were open to the public and attracted large, enthusiastic audiences. His appeal may be best summed up in Sedgwick's own words: "I cannot promise to teach you all geology, I can only fire your imaginations."[6] Darwin states that if he had attended Sedgwick's "eloquent and interesting lectures.... I should probably have become a geologist earlier than I did."[7]

Sedgwick and Henslow believed that all living forms were fixed and immutable within the biblical "kind"[8] in which they were created. Sedgwick was also a man of the cloth, having been ordained in the Church of England in 1817. He was heavily influenced by William Smith, who is famous for using fossils in sedimentary rocks to identify strata in the rocks. Smith proposed that there were only certain kinds of fossils in different rock layers and that fossils can, therefore, be used to determine the rock layer. Smith believed that these rock layers took millions of years, not the Bible's 6,000 years, to be laid down.

Smith produced the first geological maps of England and Wales, which charted the apparent strata. For the strata to occur, Sedgwick inferred that there were a series of cataclysmic events that spanned long periods of time. Sedgwick believed that the last of these catastrophes was Noah's flood, a view shared completely by Henslow.

Henslow and Sedgwick were described as Catastrophists and attracted a following, due to Sedgwick's prominence at Cambridge, one of the most prestigious universities in the world. Sedgwick also served as president of the Geological Society of London, a position of power and influence.

The Great Compromise

The prevailing earth-age theory in the 1820s was that the earth was not 6,000 years old, but hundreds of thousands or millions. For the most part, catastrophists regarded Noah's Flood as the most recent event marking the end of geological layering. In France, Georges Cuvier, the renowned French comparative anatomist and vertebrate paleontologist, published in 1813 his *Theory of the Earth,* which basically proposed multiple catastrophes, after each of which God created new creatures more complex than the prior "creation." The last such catastrophe was Noah's Flood. Cuvier called this process "progressionism" or Progressive Creationism.[9]

Adam Sedgwick's counterpart at prestigious and

influential Oxford University was William Buckland, also a popular professor of geology who advocated the long age theory. Buckland, another Church of England clergyman, asserted, as creation writer Terry Mortensen explains, "that geology was consistent with Genesis, confirmed natural religion by providing evidence of creation and God's continued providence, and proved virtually beyond refutation the fact of the global, catastrophic Noachian Flood."[10]

Darwin entered Cambridge in 1828, at a time when the assumed ancient age of the earth was exerting enormous pressure on scholars to harmonize the Genesis account with the findings of geology. Numerous reinterpretations of Genesis were offered to stretch the time frame presented in it. One, the Gap theory, suggested a chronological chasm between the first and second verses in Genesis 1. Another reinterpretation introduced at this time was the Day-Age theory, in which the six days of creation were vast ages, with the final day seven being the present time.

These reinterpretations of Genesis are still espoused by Christians who follow Buckland in assuming that Scripture must be made to accommodate geological "truth." But fossils and rocks do not have dates stamped on them. An old age for the earth is a conclusion reached by how the geological record is interpreted. So, are these interpretations correct? Is the conclusion reliable?

It is possible that Darwin, had he been taught by his

professors and mentors just how highly speculative geological dates are, might not have introduced his theory of evolution. Those who assert as fact absolute claims about the age of· the earth need to heed the words of geologist H. H. Read, who said:

> Geology, as the science of earth history, is prone to controversy. The study of history of any kind depends upon documents and records. For the history of the earth's crust, these documents are the rocks and their reading and interpretation are often difficult operations.[11]

Men such as Sedgwick, Henslow, Buckland, Cuvier, Smith, and others were reinterpreting Scripture to accommodate their geological conclusions. They were basing the geological record on the presupposition that layering of the rocks took long periods of time, rather than using the biblical chronology described clearly in Genesis. Proposals such as the Gap, Day-Age, Progressive Creationism, and other theories were offered to accommodate the data and compromise the Word of God. But does the Word of God precede man's theories or do man's thoughts, including his scientific theories, precede the Word of God? According to the Catastrophists, the latter is true. These arguments, first developed in the early 1800s, still exist today, with the result that Christian creationists are split on the question of a young or old earth.

But if we assume that the earth is old, we are also saying that death came before sin. That creates an enormous theological problem. If the fossil record is the geological account of a series of catastrophes that befell the earth before the fall of man, then death occurred before Adam. But how could everything be "good," as God stated six times in the first chapter of Genesis, if the earth contained a history of catastrophic death?

The fossil record represents death and destruction that can only come from the judgment of God on man's sin. How could there be death before sin? The fossils collected represent God's judgment and only make sense after Genesis 3, where man's fall is recorded.

The fossils used by Sedgwick to construct his model of earth history came from the great one-time catastrophic event recorded in the Bible as the Genesis Flood. These fossils were mostly shells and creatures that came from oceans, which points to a global oceanic event. The different strata layers that were observed in England and Wales did not necessarily take long periods of time to be formed. They easily could have been laid down in a short period of time during the action of the tremendous forces generated by the Genesis Flood.

These nineteenth-century geologists compromised the Word of God and fractured the foundation laid in Genesis for earth's origin—a foundation on which the rest of the Bible depends. Sedgwick and his fellow natural theologians are responsible for this crack in the foundation that would shortly be exploited by evolution and lead to a collapse of the then-dominant biblical worldview in Western culture.

After Darwin completed his theology degree in January 1831, he studied geology and attended Sedgwick's invigorating lectures, where he learned about the geological formations and the layering of rocks. After reading an account of a voyage to the Canary Islands and South America, Darwin enthusiastically decided to plan an expedition to Tenerife, the largest island in the volcanic Canary Islands. However, the trip was permanently postponed due to the sudden death of his travel partner. At this time, he was given the opportunity to join Sedgwick on a North Wales geological expedition in August 1831. For Darwin this was a great hands-on time in the field to collect rocks and see firsthand the geological formations to which Sedgwick and Henslow frequently referred.

LYELL AND THE *BEAGLE*

When Darwin returned from the North Wales expedition, he received an invitation that would change his life and redirect the course of Western civilization. Henslow had been asked to be a naturalist on the H.M.S. *Beagle*, but had to pass up this opportunity to travel around the world collecting specimens because of his other responsibilities. He offered the opportunity to his close friend, Leonard Jenyns, who also refused. Henslow then asked Darwin to go, and the recent Cambridge divinity graduate was eager for the opportunity. His father, however, strongly objected, preferring that Charles settle down and become a clergyman.

Charles' uncle, Josiah II, stood up for Charles and convinced Robert Darwin to allow his son to take the trip. At the age of 22, Charles was to travel around the world as a young naturalist on the H.M.S *Beagle*, a ship just 90 feet long. He left on the morning of December 27, 1831, with a crew of 77, and he almost immediately became seasick. His stomach was so troubled by the rough, relentless ocean that he wondered whether he would make it back to England. He did, returning five years later on October 2, 1836.

The *Beagle* was commissioned to chart the coastline of South America and run a series of chronometric readings around the world. The voyage kept close to the coastline to take measurements, which gave Darwin a unique opportunity to go onshore and collect biological and geological specimens of uncharted and unknown lands. Darwin compiled countless pages of notes, most of a geological nature. The five-year trip, in Darwin's own words, was "by far the most important event in my life and has determined my whole career."[12]

Before he left on the *Beagle*, Darwin was already

trained by his mentor, Henslow, to collect biological specimens and make geological measurements of rock formations. He was in the process of being trained as a naturalist, as he was planning his expedition to the Canary Islands. Henslow suggested Darwin take along on the *Beagle* the first volume of Charles Lyell's *Principles of Geology* as a practical factual reference when he looked at the geological formations. Henslow, however, sternly warned Darwin, "on no account to accept the views therein advocated."[13] Henslow felt that Lyell's book was "altogether wild as far as theory is concerned."[14]

Charles Lyell, a trained lawyer who was unable to practice law because of poor eyesight, had a passionate interest in geology. He took a more radical view than Henslow and Sedgwick. Lyell completely eliminated the multiple catastrophes proposed by Sedgwick and the geologists of the day, because there was no substantial evidence for these multiple worldwide catastrophes.

Past catastrophes, if they occurred, should be clearly evident in the geological column, but are not, said Lyell. He believed that present geological processes acting on the rocks today, which are measurable, are the best guide to the past. For example, the rate at which sediments fall from the oceans and rivers today is assumed to have been the same throughout earth history. According to Lyell, the processes now taking place on the earth have been taking place for millions and millions of years. This is called uniformitarianism. It basically states, "The present is key to the past."

Henslow, Sedgwick, and the British natural theologians in the 1820s all believed such thinking was heretical, because it would completely eliminate any trace of the Genesis chronology and Noah's cataclysmic deluge. The outside observer, however, may conclude that the Genesis account was already compromised by the natural theologian's belief in an old earth. They began the process of not taking God's Word seriously by reinterpreting Genesis to make it square with geological assumptions. Charles Lyell continued that process and did away with Genesis as history completely.

Sedgwick, Henslow, and others were willing to revise their understanding of what the word "day" meant in Genesis 1, but they kept a literal understanding of Noah and the worldwide Flood. Lyell just took the Flood out of the picture, and with it the Genesis creation account literally disappeared. No wonder Christians are ridiculed. If you accept Lyell's geology as scientific fact, Genesis becomes a myth. If Genesis is not true, then there is no foundation for the essential doctrines of Christianity.

The Foundation That Does Not Exist

Did scientific evidence drive Lyell to his uniformitarian conclusion? Lyell interpreted the geological record to indicate slow change over time. But this conclusion is not supported by the evidence. The geological record is incomplete—so incomplete that it was an embarrassment to Darwin and remains so to the evolutionary community today.

The late Stephen Jay Gould, a well-known evolutionary biologist, addressed this unsettling lack of evidence for the claims of Lyell and Darwin:

> ... Lyell emphasized the importance of interpreting evidence critically but not necessarily literally. The geological record, like most archives of human history, features more gaps than documents. (In a famous metaphor, later borrowed by Darwin for a crucial argument in the *Origin of Species*, Lyell compared the geological record to a book with very few pages preserved; of these pages, few lines; of the lines, few words; and of the words, few letters.) Moreover, the sources of imperfection often operate in a treacherous way, because they do not delete data at random but rather in a strongly biased fashion—thus tempting us to regard some causes as dominant merely because the evidence of their action tends to be preserved, while signs of more truly important factors may differentially disappear from the record.[15]

Gould's admission that the fossil record has more "gaps than documents" is an acknowledgement that it provides very little on which to form a scientific opinion or theory. But this is what Darwin used to reach his conclusion. The fossil record, it turns out, is the foundation that does not exist. Geologists often speak of the "geologic column" as a record in the rocks of the entire sequence of earth history. But there is, in fact, no geological column in existence anywhere on earth. As the late Dr. Henry Morris has stated, "Every local stratigraphic column has different fossil sequences from every other, so there is no obvious worldwide column that could be correlated with time."[16]

Lyell looked at the same geological record that young earth geologists do today. What one concludes from this record depends, ultimately, on what one is trained to see and interpret. Lyell's uniformitarianism depends entirely on the premise that the conditions and forces that act on the earth will be constant and continuous. His scheme has no room for catastrophes.

Creationists, by contrast, point to the fossil record and its numerous fossil graveyards as evidence of a past catastrophe. For example, the Dinosaur National Monument in Utah features a gigantic quarry wall with some 1,500 dinosaur bones exposed. This striking burial site certainly appears to be the consequence of a horrific flood with enormous energy.

Another example is the Green River Formation in western Colorado, eastern Utah, and southwestern Wyoming. It is one of the most important fossil sites in the United States, because it preserves fossils of subtropical plants and over 60 species of vertebrates. This very large site has millions of fish fossils extraordinarily well preserved. In some places you can see fine fish bones perfectly intact, without any missing or displaced. It is amazing to see these fish fossils "found packed as densely as several hundred fish per square meter of slab of rock."[17]

An abundance of fish coprolite, or feces, signify a quick burial at this site. Fossils here show evidence of fish eating fish. It is believed that these fish could have died from salinity shock, followed by suffocation. The layering in the rock clearly suggests a catastrophic event with enough energy to bury millions of fish. This certainly does not support Lyell's uniformitarian model. Today Lyell's followers are still trying to hold on with highly speculative, questionable, and suspect empirical data to support it.

Charles Lyell used suspect data and the uniformitarian assumption to reach a conclusion about the age of Niagara Falls that contradicts what biblical chronology indicates for the age of the earth. Lyell visited the Falls in 1840-41 at a time when the Falls were still quite remote and not well studied. He learned from a geologist that the waters rushing over the Falls caused the location of the Falls to recede at about three feet per year. In other words, because of erosion, Niagara Falls is moving steadily towards its source, Lake Erie, which it will, in time, reach.

For reasons that Lyell does not specify, he concluded that a "much more probable conjecture" for the rate of erosion was just one foot per year. At that rate, it would take 35,000 years for the Falls to recede the seven miles from Queenston, Ontario, (where a cliff on either side of the river indicates the original position of the Falls) to the current location. In fact, based on careful measurements, the rate of erosion is known to be four to five feet a year. If that rate has held constant over time—not likely because of the global Flood—it would have taken 7,000 to 9,000 years for the Falls to recede seven miles from Queenston to its present position.

So why did Lyell use the erosion rate of just one foot per year? Lyell presented his conclusions about Niagara Falls in editions of his *Principles of Geology*, published after 1842. Dr. John Morris, President of the Institute for Creation Research, offers an answer. "Lyell's work at Niagara accomplished its main goal," Dr. Morris stated, "that of calling Scripture into question; for biblical chronology cannot allow 35,000 years since Noah's Flood. And if Genesis is wrong, how can we trust any other portion?"[18]

Darwin ignored Henslow's counsel and used Lyell's book like a Bible to interpret his geological data. "I had brought with me the first volume of Lyell's *Principles of*

Geology," Darwin wrote, "which I studied attentively, and this book was the highest service to me in many ways."[19]

He "studied attentively," believing what he read. Darwin exchanged his mentor's belief in multiple catastrophes for Lyell's uniformitarianism. He now praised Lyell: "The great merit of the *Principles* was that it altered the whole tone of one's mind, & therefore that, when seeing a thing never seen by Lyell, one yet saw it partially through his eyes."[20]

The assumption that earth history reached back millions of years provided Darwin with what he needed to propound his idea that organisms evolve by means of imperceptibly slight changes over time. Darwin read Lyell on the *Beagle,* and when he returned to England, he immediately sought him out. They became close friends and collaborators, but Lyell adhered to fixed species, as is expressed in Volume 2 of the *Principles*. Lyell eventually came to accept Darwin's theory of evolution some time after publication of *Origin*. Darwin had an unbounded admiration for Lyell, stating once, "I always feel as if my books came half out of Lyell's brain, and that I never acknowledge this sufficiently."[21]

Darwin did, however, dedicate the second edition of *Origin* to Lyell:

> To Charles Lyell Esq., F.R.S. this second edition is dedicated with grateful pleasure—as an

> acknowledgment that the chief part of whatever scientific merit this journal and the other work of the author may possess, has been derived from studying the well-known and admirable *Principles of Geology*.[22]

It was Charles Lyell who took Darwin from catastrophism to uniformitarianism and made Darwinian evolution possible. How did this happen? When Darwin boarded the *Beagle*, he said he had two valuable books: the Bible and Lyell's *Principles*. Five years later, when he landed, he would just have one book that he studied and meditated on, and it wasn't the Bible. It was Lyell's *Principles*.

Darwin Develops "My Theory"

As a young naturalist on the H.M.S. *Beagle,* Darwin had the unique opportunity to collect specimens no man had seen before. The *Beagle* made numerous stops to chart the waters near the coastline of South America, giving Darwin the opportunity to collect specimens and make geological observations. He shipped back tons of specimens to Henslow to hold and help him classify by sending them to other experts. Darwin made exciting finds, like fossils of megatherim, a giant sloth; glyptodont; and armadillos four to five feet tall. Remains of odd-looking camels went back to England and caused a stir.

Darwin wrote copious notes of his observations and

sent them back to Henslow. By the time Darwin returned home in 1836, Henslow made certain his work was published in credible scientific journals. Henslow was well-known in the scientific circles and this man who mentored Darwin made sure the whole world would know about his favorite student. When Darwin came home, he was hailed as a hero in professional circles. He was now a well-regarded naturalist, thanks to Henslow. His notes of the voyage became valuable and gave Darwin the opportunity to become an established author and a recognized naturalist.

He narrated his experience on the voyage in chronological order and published it under the editorship of Captain Robert FitzRoy, as the *Journal of Researches into the Natural History and Geology of the countries visited during the voyage round the world of H.M.S. Beagle*. Today it is known as *The Voyage of the Beagle*.

There were other collaborative efforts with Richard Owen, who identified mammal fossils, and George Waterhouse on mammals, published in *Zoology of the Voyage of H.M.S. Beagle*. Darwin was a prolific writer who, in his lifetime, published 23 books and made over 140 contributions to books and periodicals—a great achievement for this time in history.[23]

Many presume that Darwin was not a qualified naturalist because he was a divinity student. The opposite is true. Darwin spent his life studying origins and became highly respected as a naturalist and scientist of the time. Even though he had no formal academic training in science, he took advantage of his association with some of the most esteemed scientists of the time at Cambridge. There are many who do not agree with Darwin's conclusions, on scientific and philosophical grounds (including myself), but he had a well-established reputation in the natural sciences with his hundreds of published research papers and his books.

Darwin had no interest in a career; he was well taken care of financially. He wanted, merely, "to go with the subjects to which I have joyfully determined to devote my life."[24] He entered the scientific world with a bang and came in with a collection of specimens that turned many heads. He took advantage of every opportunity and used each one to gather fossils and specimens that intrigued many, such as fossils of the giant sloth, armadillo, and exotic creatures like the Galapagos turtles and monster-like lizards.

His collection of specimens was not only intriguing to the public, but certainly attracted many outstanding scientists, including Richard Owen, as mentioned before, the famous comparative anatomist and noted creationist. This most distinguished scientist published over 600 papers himself and coauthored a paper with Darwin, as mentioned above. He was excited about Darwin's finds and gave him professional recognition, establishing him as a credible scientist.

It is interesting to note that Owen was an outspoken creationist, who had no idea that Darwin was working on his evolutionary theory when they co-published their paper. After the publication of *Origin*, Owen helped Bishop Samuel Wilberforce prepare for a debate with Thomas Huxley, Darwin's "Bulldog," in 1860. This first and famous public debate regarding Darwinian evolution was unattended by Darwin, who was sick and could not participate.

Darwin came back from his five-year voyage a changed man. The specimens he collected, particularly in the Galapagos, made a deep impression upon him. Niles Eldredge, who had the opportunity to examine Darwin's personal notebooks, saw that the observations, descriptions, and drawings were written with purpose and clear direction. It was apparent that an idea was germinating, as he started to write "my theory."[25] He was composing the transmutation theory, abandoning the idea of the immutability of species ordained by God. He started a new path of study that would definitely bring controversy and take God out of the equation.

He had to make a choice and it certainly was not going be in the direction of his favorite mentor, Henslow, or Professor Sedgwick. When Darwin published *Origin*, he sent Henslow a copy with a note: "I fear you will not approve of your pupil in this case."[26] He was right.

Henslow said, "I have told him that I cannot assent to his speculations without seeing stronger proofs."[27] Leonard Jenyns, a brother-in-law and close friend of Henslow's, summarized Henslow's reaction to *Origin*:

> In previous letters to myself, he had told me he thought there were in Darwin's book too many suppositions, too many things assumed, which might or might not be true. Moreover, the further the question was followed up towards its source, the more it was beset with difficulties which were never likely to be solved. In fact, he said, he considered an inquiry into the origin of species about as hopeless as an inquiry into the origin of evil.[28]

Darwin's *Origin* is riddled with supposition after supposition. It is void of data, charts, graphs, tables, and references to research that has been done. It basically reads as a work clouded with ambiguity and imprecision.

The same problem exists today, and those who are committed to evolution are blinded to the fact that their proposal has no substance. It continues to be "beset with difficulties." How can you test this hypothesis without seeing the process of evolution actually occur? Can you see the invisible amoeba become visible, or can you see how a monkey becomes a man? Henslow said that it is hopeless!

Adam Sedgwick, Darwin's geology professor, at one time praised Darwin to his father, saying that he "should take a place among the leading scientific men."[29] However, upon reviewing a copy of *Origin*, he was greatly disappointed, as he told Darwin in a frank letter sent on December 24, 1859:

> I have read your book with more pain than pleasure. Parts of it I admired greatly, parts I laughed at till my sides were almost sore; other parts I read with absolute sorrow, because I think them utterly false and grievously mischievous. You have deserted—after a start in that tram-road of all solid physical truth—the true method of induction, and started us in machinery as wild, I think, as Bishop Wilkins's locomotive that was to sail with us to the moon.[30]

Sedgwick asked how Darwin could come to conclusions that are based upon imaginative assumptions. Sedgwick and Henslow, and the many critics who follow to the present day, share this common objection. Where is the science in Darwinian evolution?

Henslow and Sedgwick objected to Darwin's theory because it violated the principle that God created life. According to Jenyns, Henslow objected to Darwin's theory because "it did not allow for the interposition of the

Almighty."[31]

John Stevens Henslow, the man Darwin described as a "most admirable" tutor,[32] opened the door for Darwin to become a famous naturalist. Unfortunately, what his famous student finally proposed completely contradicted Henslow's core convictions. Indeed, Darwin went beyond both the evidence and his erstwhile faith with a dangerous idea that would be fatal to mankind.

CHAPTER FOUR

DARWIN'S TREE OF LIFE

The stunning diversity of plants and animals that Darwin observed on his South American voyage made him ask this question: Why are there so many different types of turtles, lizards, birds, insects, etc., to be observed across the world? And why do all these different kinds of animals and plants all seem to be perfectly adapted to their geographic environment? As he toured the Galapagos Islands, he struggled to understand why these islands had such unique creatures, which nonetheless roughly corresponded to animals observed on the continent.

He discovered in South America the fossils of giant extinct animals, such as an 18-foot-tall sloth and a five-foot-long armadillo. These exciting finds led to more questions about why the body plan of these extinct animals so closely corresponded, except for size, to their contemporary counterparts. Other fossil finds, such as the unique *Macrauchenia,* which resembles a camel, and *Capybaras*, large rodents similar in appearance to their living rodent cousins, raised the question of why these animals became extinct.

Darwin described his intriguing observations:

> During the voyage of the *Beagle* I had been deeply impressed by discovering in the Pampean formation great fossil animals covered with armour like that on the existing armadillos; secondly, by the manner in which closely allied animals replace one another in proceeding southwards over the Continent; and thirdly, by the South American character of most of the productions of the Galapagos archipelago, and more especially by the manner in which they differ slightly on each island of the group; none of the islands appearing to be very ancient in a geological sense.[1]

The armadillos Darwin was referring to are the extinct glyptodonts, which were quite unlike any armadillo on planet earth today. Growing to the size of a small car, they were covered in impenetrable shell armor up to five centimeters thick. From his observations, Darwin assumed that the glyptodont was an ancient ancestor of the armadillo and evidence of change over time.

The conclusion Darwin came to from all this was that *change is taking place through time,* which, by definition, is evolution. "It was evident," Darwin wrote in his *Autobiography*, "that such facts as these, as well as many others, could only be explained on the supposition that species gradually become modified; and the subject haunted me."[2]

These modifications that "haunted" Darwin can be best explained not as evolution, but as variation within the sloth or armadillo kind. Could the huge size of these fossils be due to gigantism? The tallest man in medical history measured 8 feet 11 inches, while the smallest tribe of pygmies, from Zaire, Africa, can be as small as 4 feet 1 inch.[3] Despite the size range, both are still human.

What about modifications in the plumage or beak size of birds? Is this evolution? How would we know? If we do not know how to define a species, how can we know when enough change takes place to constitute "transmutation," which is Darwin's term for evolution from one species into another?

The creationist perspective is that modifications such as beak size are but variation within a kind, not the creation of a new kind. The point of controversy is whether modifications may be so extensive that so-called microevolution (variations within a kind) leaps beyond the species boundary and at some point becomes macroevolution. But of course, since "species" is not defined, it is impossible to say when the species barrier is broken. Because of this vague definition, small changes or large changes can be said to produce a new "species."

When Darwin stepped off the *Beagle* in October 1836, he was obsessed with explaining how living systems could gradually change over time. His assumption that they do runs contrary to the biblical idea of a fixed "kind" established in Genesis 1. The idea of continuity within each kind is explicit in Genesis 1:12, which states, "And the earth brought forth grass, the herb that yields seed according to its kind, and the tree that yields fruit, whose seed is in itself according to its kind." The "kind" here is dictated from the seed, which is controlled by DNA.

Darwin believed that the modifications he observed did not occur quickly, but required great stretches of time. Lyell's uniformitarianism, which denied catastrophism and said the same geological processes observed now were present in epochs past, made it possible for the process of natural selection to operate without catastrophic interruption over those long ages.

Darwin had a great deal of work ahead of him when he returned to England. He had an extensive collection of fossils and geological specimens that had to be identified and classified. He took up residence at Cambridge and began the arduous task of going through his enormous collection and developing his ideas about the transmutation of species.

At this time he contracted an illness that remained undiagnosed and a mystery for the rest of his life. He suffered a constant upset stomach that caused vomiting, insomnia, headaches, and palpitations of the heart. He had enjoyed good health when he was on the *Beagle*, but his son Francis reported that, for his father, "nights were generally bad, and he often lay awake or sat up in bed for hours, suffering much discomfort."[4] Darwin wrote to FitzRoy, the captain of the *Beagle*, stating, "I have nothing to wish for excepting stronger health."[5] It has been said that he might have contracted a tropical disease, such as Chagas disease, while collecting his many specimens.

Although Darwin had a wonderful opportunity to observe a great variety of different kinds of living systems and geological formations in South America and the Galapagos Islands, his five-year voyage was, nonetheless, limited in space and time. He could not investigate every place on earth and had sharp limits placed on the amount of time he spent in each place.

Captain FitzRoy's mission to chart the waters around

South America took priority over Darwin's natural history work. Darwin was not paid for his "collecting," but was, in fact, a paying guest. The ship did not wait for him to finish gathering samples in the woods or in the mountains. When Captain FitzRoy gave the order to launch, Darwin had to be there.

He not only had limits on his time, but he also had no control over where his collection activity took place. He was entirely dependent on where the ship anchored and, as a result, the data he collected had no scientific continuity or purpose. Although Darwin's collection proved to be quite interesting, it was not the result of a careful systematic search, but the happenstance result of what was to be found at the locations where FitzRoy stopped.

Darwin's task on the voyage of the *Beagle* was to do natural history research, but natural history is a vast area of study which encompasses knowledge about insects, birds, geology, plants, minerals, chemistry, and so on. To get a complete and accurate picture of the earth and its history is an extremely challenging undertaking. It requires an abundant amount of data from many disciplines of science. The data requirement is exponentially greater if one assumes that the earth is not thousands of years old, but millions or billions of years.

Darwin could not claim expertise in all the scientific disciplines listed above and was therefore constrained by the limits of his own knowledge. His observations and the data he collected are obviously but a fraction of what can be known about nature. If life on earth goes back millions of years, the earth should contain a near-infinite number of data points in both the fossil and geological records. This rich data trove should, as with a mega-pixel camera, provide more than enough information to provide a high-resolution snapshot of the past. But that is not the case. The actual earth history record is embarrassingly slim and incomplete. We do not have a clear picture of the past.

There is overwhelming evidence to suggest that the missing data points from the past are the result of a worldwide flood, as described in Genesis. This global water event unleashed tremendous destructive forces clearly seen in the fossil and rock record. Darwin refused to see this, but cast his lot with Lyell in assuming a uniformitarian earth history and gradual change through long periods of time.

Darwin collected a great number of fossils and wrote copious notes, but his research constitutes a minute part of earth history. Darwin went on to make imaginative speculations without the requisite scientific data. He did not have an evidential foundation to reach his conclusions and to change the world's thinking about natural history and the Creator.

Like every other human being on this planet, Darwin was finite and subject to error. The state of paleontological knowledge is under constant refinement, as more and more bones are collected and more pieces of the puzzle are found. This is true of Darwin and his *Beagle* voyage collection. Darwin was wrong, for example, in thinking that giant extinct animals like sloths and glyptodonts were unique to South America. In fact, they are also found in many different areas of North America.

Darwin thought that these creatures were replaced via evolution by others who were better built for survival. He inferred that these sloths evolved from other creatures, like the anteater, over millions and millions of years. This inference cannot be observed and is a source of error. In his preface to *Origin*, Darwin states,

> I was much struck with certain facts in the distribution of the organic beings inhabiting South America, and in the geological relations of the present to the past inhabitants of that continent. These facts, as will be seen in the latter chapters of this volume, seemed to throw some light on the origin of species—that mystery of mysteries, as it has been called by one of our greatest philosophers.[6]

These "facts" include his mistaken assumptions that later had to be changed. So, how reliable are Darwin's assertions that are based on inferences that can't be tested? On what basis can he assert that his "facts" will bring light to that "mystery of mysteries"—the origin of species?

Darwin's Finches

Darwin carefully observed the beaks on 13 different varieties of Galapagos Islands finches. The beak and body size of each seemed to be perfectly fitted for the particular island it inhabited. These "Darwin Finches" demonstrated, he claimed, how a species can change over time. Darwin took specimens back to England to observe and share with other naturalists. Eldredge claims that these finches are one of the most compelling arguments that led Darwin to his theory of evolution. He quotes from Darwin's narrative in the *Voyage of the Beagle* to emphasize the point that modification really means evolutionary change:

> Seeing this graduation and diversity of structure in one small, intimately related group of birds, one might really fancy that from an original paucity of birds in this archipelago, one species had been taken and modified for different ends.[7]

The finches really do not demonstrate evolution, but are great examples of variation within a kind. A Princeton zoologist concluded after years of study that the Galapagos finch population did change with environmental conditions.[8] A great example of this was observed during drought years, when finches with larger, deeper beaks grew in number. This was because the traditional finch food supply of smaller seeds was soon depleted, which favored finches with beaks built to succeed when the food supply changed.

But what happens when the rain returns? Does beak size change again? The answer is yes. What Darwin observed was an incomplete picture. It was not possible for Darwin to get an accurate assessment of how finch beaks varied based on just one visit. In addition, finch features change very quickly, because they can interbreed. Peter and Rosemary Grant, who have lived and worked on the Galapagos Islands for over twenty years, witnessed many pairings of "different species" of finches. It is hard to determine which finch came from which finch if they interbreed and produce fertile young. The Grants predicted that the finches would change not in millions of years, but in 200 years:

> While that is not very surprising, nor profound, the speed at which these changes took place was most interesting. At that *observed* rate, Grant estimates, it would take only 1,200 years to transform the medium ground finch into the cactus finch, for example. To convert it into the more similar large ground finch would take only some 200 years.[9]

The difficulty that exists in simply identifying finches makes these elusive creatures a very weak reed on which to base far-reaching conclusions about evolution. Darwin himself misclassified finches. When he came back from the *Beagle* voyage, he had the finches reevaluated by John Gould, an expert in bird identification, and he found that the finches were part of the same group, not different groups, as he had thought.[10]

David Rothman, at the Department of the Biological Sciences at Rochester Institute of Technology, states that, "Identification of finches can be extremely intimidating."[11] According to Dr. Rothman:

> Despite the fact that they intrigued Darwin, they are far too complex a group of animals for Darwin to have understood. Darwin's finches share similar size, coloration, and habits. Their salient difference is in the size and shape of their beak. However, beak shapes can be very variable, and the size and shape in one individual can overlap into the range of another species.[12]

Darwin and those that wave the evolutionary flag need to know that a finch is still a finch. Beak size, body size, even plumage may well change—and are expected to do so, due to variation within a kind. Although Darwin used finches to illustrate his claim that change occurs over time, he chose a bird that was too complex for him to identify. The shape and size of the finch beak are too difficult to readily identify. In spite of these difficulties, finches still appear in textbooks as evidence of evolution.

Darwin's Tree and its Soil

Darwin began his "Transmutations Notebooks" in 1837 with a reference to his grandfather's work, *Zoonomia*, which promoted the idea of evolution. The "tree of life" shows up in this notebook when Darwin states, in cryptic fashion, "changes not result of will of animal, but law of adaptation as much as acid and alkali—organized beings represent a tree, *irregularly branched,* some branches far more branched—hence Genera—many terminal buds dying, as new ones generated."[13]

Under the words, "I think," Darwin sketches his primitive vision of the tree of life. Here, for the first time, Darwin outlines in rude form how he thinks molecules became men over millions of years by common descent. The trunk of this tree is, he believes, the common source from which all living forms arose and is rooted in simple, inert molecules.

Today Darwin's tree of life is diagrammed in a more elaborate fashion, with straight lines that branch from a common line, but the concept remains the same: Living things change over time and form new living things that bear a body plan that resembles their antecedent. This tree of life eliminates the need for an outside source, a benevolent Creator who spoke His creation into existence. This infinite Source, from which all creation obtains energy, direction, and most importantly, purpose, is discarded in favor of Darwin's crudely drawn tree of life.

Just before he introduced his tree of life in his notebook on transmutation, Darwin, in what Eldredge describes as a "watershed" passage, moves from God the Creator to natural laws as the means by which nature came to be. Darwin noted that just as the once widely held idea that God orders the planets by special providence has been superseded by the knowledge that the heavenly bodies move by the laws of astronomy, the idea that animals are His special creation should give way to what Darwin in his notebook calls "the fixed laws of generation." Darwin states:

> Astronomers might formerly have said that God ordered each planet to move in its particular destiny—in same manner God orders each animal created with certain form in certain country, but how much more simple and sublime power let attraction act according to certain laws such are inevitable consequences let animal be created, then by the fixed laws of generation, such will be their successors....[14]

According to Eldredge, "Darwin himself becomes rightly enamored of this grand view, expanding the range of scientific law from physical laws of motion applied to the external universe to the history of life on earth."[15] Having replaced the personal creator God, Darwin is ready to substitute "fixed laws of generation" which become the soil in which the tree of life takes root. God is no longer a part of the creative process. Eldredge offers his praise:

> The result: evolution, the idea that all life on earth, from time immemorial, is descended from a single common ancestor. A result so powerful that it ranks among the great ideas in the history of the Western world.[16]

But how does life move upward on the tree of life without a benevolent Creator? What force compels life to arise from lifeless molecules and atoms? Darwin recognized that his tree needed a mechanism to move it forward. He was still missing a means by which the tree of life could be brought to bud and sprout.

The answer came in 1838 when Darwin read an essay written by the Rev. Thomas Robert Malthus, an English demographer and political economist. His *Essay on the Principle of Population* purported to show how the growth of food resources must always lag behind human population growth. For Malthus, "the power of population is indefinitely greater than the power in the earth to produce subsistence for man. Population, when unchecked, increases in a geometrical ratio. Subsistence increases only in an arithmetical ratio."[17]

While Malthus, whose essay was published in 1798, did not address evolution, his ideas proved useful to Darwin in winning acceptance from the scientific community for the concept of evolution. Malthus introduced the phrase, "struggle for existence" to describe the process of obtaining adequate food from limited resources. He focused not on animals and plants, but the ability of an economy to sustain a growing population and predicted that, left unchecked, the human population would exceed the capacity of man to produce adequate food.

The grim scenario Malthus predicted has not been realized. Food production has increased geometrically through new technology and harvesting techniques. What Malthus and others, including more recently, Dr. Paul Ehrlich, author of *The Population Bomb*, have failed to realize is that people are not mere consumers of resources. Instead, the minds and hands of people are themselves resources, endowed by God with energy, industry, and the ability to innovate in a manner that increases the sum of resources available to mankind.

For Malthus, the struggle for existence is hindered by vice and misery. He believed that accidents, old age, war, pestilence, famine, infanticide, murder, contraception, and homosexuality would check population growth. Moral restraint—delaying marriage and sexual abstinence before marriage—could also, he believed, help stem population growth. "By moral restraint," he wrote, "I would be understood to mean a restraint from marriage, from prudential motives, with a conduct strictly moral during the period of this restraint;...."[18] Malthus limited his counsel to the poor and members of the working class, for which he received much criticism.

Malthus believed that mankind's misery and vice are the consequence of the Fall and the curse that followed. He issued a moral plea, urging men to remove the evil from their hearts and to find hope in the Gospel of Jesus Christ—a message I wish Darwin had taken to heart.

Darwin applied to nature what Malthus proposed for an economy. He took the Malthusian "struggle for existence" and proposed, by analogy, the "survival of the fittest," or "natural selection," as the mechanism by which new life forms came into existence in nature. Darwin explained in *Origin* that after reading Malthus' *Population*, "it at once struck me that under these circumstances, favourable variations would tend to be preserved and unfavourable ones to be destroyed. The result of this would be the formation of a new species. Here, then, I had at last got hold of a theory by which to work."[19]

Darwin defines natural selection in *Origin* as, "the

preservation of favorable variations and the rejection of injurious variations."[20] Natural selection was a fight to the end for survival; those individuals who survived had the beneficial traits to continue their own stock in an upward direction.

Darwin's assumption that evolution is moving in an upward direction is not, however, borrowed from Malthus. For Darwin, nature is much superior to humanity in working out the struggle for existence. Darwin writes in *Origin* that:

> Man can act only on external and visible characters: Nature, if I may be allowed to personify the natural preservation or survival of the fittest, cares nothing for appearances, except in so far as they are useful to any being. She can act on every internal organ, on every shade of constitutional difference, on the whole machinery of life.... Can we wonder, then, that Nature's productions should be far "truer" in character than man's productions; that they should be infinitely better adapted to the most complex conditions of life, and should plainly bear the stamp of far higher workmanship?[21]

Nature, which is "of far higher workmanship than man," replaces the benevolent Creator as the agent of creation. Malthus, while misguided about population and food resources, wrote to both warn and offer hope to sinful man. Darwin appropriates Malthus not to give hope, but to enthrone nature as "god." Darwin outlines a dismal struggle for existence, which ultimately leads to death. He offers not a tree of life, but of death.

British naturalist Alfred Wallace read the Malthus essay in 1846, eight years after Darwin, and like Darwin, used Malthus to develop a theory of evolution exactly like Darwin's. He wrote:

> The most interesting coincidence in the matter, I think, is that I as well as Darwin was led to the theory itself through Malthus—in my case it was his elaborate account of the action of "preventive checks" in keeping down the population of savage races to a tolerably fixed but scanty number. This had strongly impressed me, and it suddenly flashed upon me that all animals are necessarily thus kept down—"the struggle for existence"—while variations, on which I was thinking, must necessarily often be beneficial, and would then cause those varieties to increase while the injurious variations diminished.[22]

Made by Malthus a believer in natural selection, Wallace set off on a specimen-collecting tour of the

Amazon from 1848 to 1852 to prove his theory. His ship caught fire on the way home to Britain and sank, taking with it many of his specimens. This unfortunate incident did not stop him. Two years later he went on an eight-year collection expedition to the East Indies. It was during this time that he developed his own theory of evolution, using the same mechanism, natural selection, settled upon by Darwin.

Wallace sent an essay on his theory to Darwin in 1858, with a request for advice. The paper stunned Darwin, because it was nearly identical to his own theory that he had been working on for over 21 years. He had delayed publishing his theory in order to continue its refinement and to avoid the controversy it would provoke in Victorian England.

The work of Wallace prompted Darwin to publish *Origin* and to coauthor with Wallace a formal paper on their theory before the Linnean Society on July 1, 1858. Wallace is considered by many to be a co-developer of natural selection.

When Darwin did publish, he found a ready audience for his ideas. His first edition of *Origin*, published in 1859, sold out the first day and was republished in six subsequent editions.

Darwin's "Damnable Doctrine"

Darwin believed in a divine Creator when he walked

onto the *Beagle* in 1831. Five years later, he rejected the idea that God created the world and that evidence of His intelligent design is inscribed in nature. While he took William Paley's divine watchmaker thesis on trust, having been "charmed and convinced by the long line of argumentation,"[23] by journey's end he rejected Paley's brilliant and compelling arguments published in his *Natural Theology; Evidences of the Existence and Attributes of the Deity. Collected from the Appearances of Nature.*

The premise of Paley's *Natural Theology* is that nature affords innumerable instances of astonishing design. Paley noted that a watch and a stone on the ground will be understood by an observer to signal, respectively, an object formed by intelligence and a naturally occurring object. The watch represents complexity, which is demonstrated by its design and purpose. Paley's examples of design include the human frame, bones, muscles, vessels, comparative anatomy, and so on.

He presented a nineteenth century version of the argument for intelligent design, an idea which today is sparking so much controversy. We know from molecular biology that there is an unbelievable amount of complexity in the simplest life forms. Not only that, but the extraordinarily complex "simple cell" contains mechanisms within it that work together in harmony to achieve a common purpose. This design is not the outcome of natural selection—the "far higher workmanship" of nature—but is the handiwork of the Deity.

Darwin's real problem was not in Paley's *Natural Theology*. It was in Darwin himself. As Paley concludes in the last chapter of *Natural Theology*:

> The existence and character of the Deity, is, in every view, the most interesting of all human speculations. The true theist will be the first to listen to *any* credible communication of Divine knowledge. Nothing which he has learned from Natural Theology will diminish his desire of further instruction, or his disposition to receive it with humility and thankfulness. He wishes for light: he rejoices in light. His inward veneration of this great Being, will incline him to attend with the utmost seriousness, not only to all that can be discovered concerning him by researches into nature, but to all that is taught by a revelation, which gives reasonable proof of having proceeded from him.[24]

While Darwin collected a large number of specimens for study, he was unable to identify many of them because of his limitations as a naturalist. In addition, as noted earlier, Darwin's findings represented just a narrow slice of animals and plants to be found on the globe. He collected a limited number of specimens over a limited time in a limited area of the world. Those limits, however, did not place a boundary on the conclusions Darwin drew from them.

Darwin's limited data set may explain his frequent use of the subjunctive mood in *Origin*. His work takes on a speculative tone because of the recurrent use of phrases such as the following, taken from *Origin*: "we may suppose," "if we suppose," "we have only to suppose," "we suppose," "let us suppose," "let us now suppose," and "now if we suppose." Darwin made huge inferences and far-reaching assumptions that led to serious error.

In the meantime, Paley's "watch," dismissed by Darwin, has today become even more complex, as our capacity to peer into the structure of nature has grown in sophistication. Darwin reached the wrong conclusion about the existence of God and so began the modern war waged by Darwin and his progeny against the Creator.

Darwin believed that nature's creator is *natural selection*—a phenomenon he could not observe, but which was essential for the theory of evolution to work. Having rejected Paley's argument for design, Darwin effectively embraced a new religion. He stated:

> The old argument of design in nature, as given by Paley, which formerly seemed to me so conclusive, fails, now that the law of natural selection had been discovered. We can no longer argue

> that, for instance, the beautiful hinge of a bivalve shell must have been made by an intelligent being, like the hinge of a door by man. There seems to be no more design in the variability of organic beings and in the action of natural selection than in the course which the wind blows. Everything in nature is the result of fixed laws.[25]

Darwin claims in his *Autobiography* that he was an "orthodox" Christian when he went on the *Beagle* but had a change thereafter. Despite that change, he was honored with burial in Westminster Abbey when he died in 1882 at age 74. Darwin did not often speak openly about his religious beliefs, which he judged a private matter, but he elaborated on this topic in a letter written three years before his death. In a letter to a Mr. J. Fordyce, he described his belief about God:

> What my own views may be is a question of no consequence to any one but myself. But, as you ask, I may state that my judgment often fluctuates.... In my most extreme fluctuations I have never been an Atheist in the sense of denying the existence of a God. I think that generally (and more and more as I grow older), but not always, that an Agnostic would be the

more correct description of my state of mind.[26]

Darwin was not certain how to label himself, but settled on agnostic, one who says he does not know if God exists. The bottom line is that Darwin could not say affirmatively that there was a God. A persistent rumor asserts that Darwin recanted at his death bed and confessed Jesus Christ as his Savior. These claims have been refuted by his daughter Henrietta. His son Francis wrote that his father sensed that death was approaching the evening before he passed away and said, "I am not the least afraid to die."[27]

Darwin was taught the Scriptures as part of his daily lessons as a youth. He studied theology at Cambridge in divinity school, but there is no evidence that he had a life-changing experience with Christ or that he even participated in church activity. Church attendance is hardly mentioned in his autobiography. It appears that Darwin did attend Sunday worship with his wife until the death of his ten-year-old daughter, Annie. Darwin loved her dearly and became bitter towards God, whom he blamed for allowing death and suffering.

Darwin made it plain that he once believed in God and the immortality of the soul but later likened his thoughts about God to mere sentiment, on the order of how one feels when listening to music. While he once found himself impressed with the idea of God when observing a "grand scene" such as the Amazon rainforest, he stated later:

> I cannot see that such inward convictions and feelings are of any weight as evidence of what really exists. The state of mind which grand scenes formerly excited in me, and which was intimately connected with a belief in God, did not essentially differ from that which is often called the sense of sublimity; and however difficult it may be to explain the genesis of this sense, it can hardly be advanced as an argument for the existence of God, any more than the powerful though vague and similar feelings excited by music.[28]

The edition of Darwin's *Autobiography* compiled by his son Francis is replete with details about Darwin's interaction with his children, his walks in the garden, his physical description, clothes he liked to wear, what he drank, his normal day, and the dogs with which he loved to play—including their names. This blizzard of detail includes precious little mention of Darwin's religious life or his devotion to church or the Word of God. The account of his youth and early manhood offers little detail about his religious convictions or activities and, after the voyage of the *Beagle*, his irreligion was firmly in place. He disregarded God and lived as if He did not exist.

That said, it is fascinating to read Darwin's own assertion of orthodoxy at the time he boarded the *Beagle*.

> During these two years I was led to think much about religion. Whilst on board the *Beagle* I was quite orthodox.... But I had gradually come, by this time, to see that the Old Testament from its manifestly false history of the world, with the Tower of Babel, the rainbow as a sign, etc., etc., and from its attributing to God the feelings of a revengeful tyrant, was no more to be trusted than the sacred books of the Hindoos, or the beliefs of any barbarian.[29]

When Darwin boarded the *Beagle* he carried a Bible and said he believed every word. He even quoted the Bible as the "unanswerable authority on some point of morality"[30] to several officers. Darwin was a writer and thinker who, on the voyage, had long periods of down time to meditate and filter all his observations and thoughts on God's Word. There is no Christian zeal recorded in any of his notes and correspondence. It is apparent he not only questioned the Bible but became particularly critical of Genesis, the most attacked book in the Bible.

Darwin's theory of evolution was totally contrary to the literal account of Genesis and led him to criticism and disbelief. Six years before his death, Darwin wrote about the process by which his former faith eroded:

> But I found it more and more difficult, with free scope given to my imagination, to invent evidence which would suffice to convince me. Thus disbelief crept over me at a very slow rate, but was at last complete. The rate was so slow that I felt no distress, and have never since doubted even for a single second that my conclusion was correct. I can indeed hardly see how anyone ought to wish Christianity to be true; for if so the plain language of the text seems to show that the men who do not believe, and this would include my Father, Brother and almost all my best friends, will be everlastingly punished. And this is a damnable doctrine.[31]

Much of the above quote was edited out of the original autobiography, and perhaps for good reason. Darwin was known as a kind, peaceful gentleman. Darwin's wife wanted to protect the image of her husband, whom she loved dearly, so she requested certain passages be removed. The revelation that Darwin took such a dim view of Christianity would have made her husband somewhat notorious to the British public. The slightly less controversial stance, that he was a freethinker who took issue with Christianity, was the picture that Emma wanted to present to the public.

She knew his theories and did not agree with him in many cases. Shortly after they married in 1839, Emma expressed concern about her husband's apparent skepticism. She urged her husband always "to give your whole attention to both sides of the question" and warned that "there is a danger in giving up revelation [the Bible]...."[32]

In another letter from Emma, written before 1861, she reaches out in deep concern for her husband. She writes in tender words to encourage him and to point to God as a source of comfort and hope. She begins this short letter excited, because he has reached a major goal with the publication of *Origin*, and adds this appeal:

> ... I find the only relief to my own mind is to take it as from God's hand and to try to believe that all suffering and illness is meant to help us to exalt our minds and to look forward with hope to a future state. When I see your patience, deep compassion for others, self command and above all gratitude for the smallest thing done to help you, I cannot help longing that these precious feelings should be offered to Heaven for the sake of your daily happiness. But I find it difficult enough in my own case. I often think of the words, "Thou shalt keep him in perfect peace whose mind is stayed on thee."[33]

Emma was a devoted wife, willing to be at her husband's side no matter what. She reached out to her husband so that he could have peace with God and not be angry. The man of patience, compassion, self-command, and gratitude described by Emma had, I believe, a war going on inside. Darwin was angry because of the judgment that will come from God. His passion overflowed in his stinging reference to eternal judgment as a "damnable doctrine."

Darwin was asking, "How can God condemn me, my father, my grandfather, and almost all my friends? Who is this God that can judge me?" Something of his inner soul comes into view here. He was unwilling to acknowledge the Creator and crafted a theory to replace Him with evolution. Darwin knew the Scriptures, but did not embrace the truth that there was One who suffered condemnation, though He was perfectly innocent. This One, Jesus Christ, did not curse his Creator, but became a bloody sacrifice for mankind and rose from the dead so that we who believe may have the wholly unearned and undeserved gift of eternal life.

Darwin did not want to give God glory and thanksgiving. As a result, man did not gain, but lost his dignity and his stature as God's creature, and became subject to God's wrath (Romans 1:18).

CHAPTER FIVE

Preservation of the "Favoured Races"

Darwin left the top branch of his tree of life empty when he published his *Origin of Species* in 1859. He did not address the question of man's evolution, telling his readers only that in the future "light will be thrown on the origin of man and his history."[1]

Not that Darwin did not make the evolution of man a central focus of his cogitations. "As soon as I had become, in the year 1837 or 1838, convinced that species were mutable productions," Darwin wrote in his *Autobiography*,

"I could not avoid the belief that man must come under the same law."[2] In other words, Darwin had reached the conclusion that man is not a distinct creation of God but, just like the rest of the animal kingdom, he is an outcome of the inexorable law of natural selection at work over millions of years. Man is, in Darwin's view, not a little lower than the angels, but just a bit above the apes—a view with ominous implications for man.

Indeed, Darwin believed that primates were man's closest relatives. The magic of natural selection working through gradual modifications over eons of time had, he believed, produced men from apes. Darwin first sketched his idea about the evolution of man in 1838 in his notebook "M." Here he included man in the tree of life and made him, like the rest of nature, a product of the implacable, sovereign force of natural selection. There was, after all, no reason to exempt man from this universal law. He had no special place and was no different from every other member of the animal kingdom.

During this time, Darwin made many visits to Regent's Park Zoo in London, where he carefully observed Jenny, a clothed orangutan. Jenny wore a little girl's white dress and received much attention at the zoo. Among the observers, most taking jocular delight at Jenny, was a serious gentleman intently taking careful notes. Darwin's observations were aimed at noting similarities in physical features and behavior patterns between simians and humans. Darwin, in fact, observed his first two children in similar careful fashion to compare their behavior to that of Jenny. All this was based on the assumption that man descended from monkeys and, therefore, holds no special place in the natural order.

Darwin understood that his conclusions about man's descent from apes would provoke a reaction. He knew that opposition would come from clergy and members of British society, who would be repelled by the idea that grunting, hairy apes were the distant cousins of cultured man. Today, even after 150 years of indoctrination via public education, museums, the popular media, and scientific literature, the ape-to-man scenario continues to engender opposition. The American people are not comfortable with monkeys in their family tree.

Polling reflects this. A 2006 Gallup poll found that 46 percent of Americans still believe that God created man in his present form within the last 10,000 years. Some 36 percent think that man evolved over millions of years in a process guided by God. Only 13 percent think evolution happened without God's involvement.[3]

Knowing that a backlash of public criticism would follow his inclusion of man in his grand evolutionary scheme, Darwin earnestly collected data and took copious notes in preparation for the publication of his ape-to-man theory. So much material was collected between 1838 and 1871 that he presented his theory in two volumes. The first volume, *The Descent of Man and Selection in Relation to Sex*, published in 1871, sought to demonstrate man's animal ancestry. The sequel, *The Expression of Emotions in Man and Animals*, published in 1872, sought likewise to show the link between animals and man by the study of their common emotions.

The Expression of Emotions was the more easy to read and popular of the two—in part because of the abundance of pictures showing the facial expressions of people and animals. Darwin had over 100 pictures or illustrations in this volume, which by itself made his book a remarkable publishing event, since photography was then at its infant stage and very expensive.[4] Darwin wanted to demonstrate that emotions were not unique to man but were shared by other members of the animal kingdom. Darwin suggested that the gamut of human emotions—anger, embarrassment, joy, sorrow, mirth, pensiveness—were evidence of evolution since, as he attempted to show, they had their more primitive expression in animals just a few rungs below on the evolutionary ladder.

The Expression of Emotions in Man and Animals, Darwin's last treatise on evolution, failed miserably to make the case for the evolution of man. The book did capture public attention with its images of a man undergoing electric shock, the cries and smiles of a baby in close proximity to an adult actor, an insane lady stirring in anger, and other illustrations. Book sales were enormous. *The*

Expression of Emotions sold 9,000 copies in four months but, despite brisk sales, the book failed to demonstrate that humans are just a bit further evolved in their range of emotional expression.

In fact, primates, the closest in Darwin's view to humans, have a very limited range of facial expressions. Chimpanzees are the most facially expressive of all primates. They have a great deal of muscular control in their lips and can put on comical faces, which make them a natural favorite at zoos. But these facial expressions have a narrow scope, centered on the lips, and are not true expressions of emotions. Mark Cosgrove, author of *The Amazing Body Human*, writes that, "Primate messages are largely limited to 'I am here,' 'I'm excited,' 'I'm afraid,' 'Get away!'"[5]

A second significant distinction between men and animals has to do with the source from which facial expressions arise. Darwin and his followers have completely neglected the fact that animal facial expressions are controlled by the limbic system, while human facial expressions are controlled by the cerebral cortex. The limbic system controls reflex reactions while the cerebral cortex is the source of abstract thought.

Because human facial expression is directly connected to the brain, the face is a rich palette able to display a broad range of emotional communications. No animal even comes close to the emotional plasticity of the human face.

There is a physiological reason for this. The complicated and far-reaching range of facial expressions that play across the human face is due to the 28 subcutaneous facial muscles that make possible a quarter million different facial expressions. No animal comes close.[6]

With these two volumes, *Descent* and *Emotions*, Darwin completed his work on evolution. *On the Origin of Species* (1859) is by far Darwin's most important work, but it was *The Descent of Man* that had the greatest implication for humanity. Darwin states in his introduction to *Descent* that his purpose was to inquire 1) as to whether "man, like every other species, is descended from some pre-existing form," 2) what was "the manner of his development," and 3) what is "the value of the differences between the so-called races of man."[7]

The assumption that man, just like the rest of the animal kingdom, is the product of evolution, leads necessarily to speculation about the "value of the differences" between groups of humans. If man evolved, then the differing traits exhibited by the "races" of men become a means to sort and classify humanity. The effort to place a value on differences between people groups is a logical consequence of Darwin's decision to reject the Creator—the One who made man in His image and gave him a "crown of honor and glory" (Psalm 8:5). Having rejected the Creator, Darwin gave no credence to God's estimate of man. In so doing, he cracked opened the door for horrors to come, as his disciples, decades later, made draconian value judgments on other "races" of men.

Darwin's devaluation of man depended, in part, on his effort to prove that animals were similar to humans in intelligence. He claimed that the distinction between man and animals was one of degree, not kind, and asserted that there is no fundamental difference between man and higher mammals in their mental faculties. Such an assumption cannot be supported by scientific literature, because "practically nothing is known about the way information processing actually takes place in the brain. Nobody knows how the perceived semantic information is derived from incoming electrical signals."[8] We do not know how certain memories are stored and have no clue how new ideas are formed. "The little we know about the functions of various parts of the brain has essentially been observed by observing changes resulting from brain damage or tumors."[9] Measuring the neuron response of an animal gives tiny insights to the human brain but does not provide a comprehensive explanation of cognitive function.

The Biological Ancestry of Man

Darwin had difficulty in tracing out the ancestry of the human race.[10] He had very little experience in human paleontology and had to borrow, as he acknowledges in *Descent*, from Ernst Haeckel's *Natürliche Schäpfungsgeshichte*,or *Natural History of Creation*, a work in which Haeckel, as Darwin put it, "fully discusses the genealogy of man."[11]

Haeckel is an extremely controversial figure. A trained German physician and devout evolutionist, he used faked data in his embryo research to prove evolution. Haeckel falsified drawings of various vertebrate embryos (chick, hog, calf, rabbit, etc.) in an attempt to show they are all almost identical in the earliest stage of development, after which they differentiate into unique animals. Haeckel called this a "biogenetic law" and adopted the phrase "ontogeny recapitulates phylogeny" to describe how the stages of evolution are, he claimed, repeated during the development of an embryo. This "law" of Haeckel's has been largely rejected by all sides in the evolution debate.

Jonathan Wells, author of *Icons of Evolution*, notes that the alleged similarity between embryos at the outset of life was to Darwin convincing evidence for his theory. Darwin believed that "community in embryonic structure reveals community of descent."[12] In other words, embryos look alike at the earliest stage of development because they are evolved from a common ancestor.

Darwin wrote in *Origin* that he found it "probable from what we know of the embryos of mammals, birds, fishes and reptiles, that these animals are the modified descendants of some ancient progenitor...."[13] This embryo similarity was, for Darwin, "by far the strongest single class of facts in favor of" his theory.[14] But, of course, the Haeckel illustrations on which Darwin reached that conclusion were faked. Vertebrate embryos are *not* similar in appearance, as Haeckel made them to appear.

Just as Darwin looked to Haeckel for expertise in embryology, he also used the German physician as an authority on race. Again, Haeckel proved a source of faulty and false information. In his 1868 book, *Natural History of Creation,* Haeckel includes illustrations of 12 facial profiles—six human, six simian—arranged by number according to their ranking on the evolutionary scale. Number one is a European, after which the following profiles are shown in "descending" order: an East Asian, a Fuegian, an Australian, a black African, and a Tasmanian. After the Tasmanian came a gorilla and other apes.

Haeckel had no empirical evidence on which to base his ordering of men and apes, which showed the "lowest" man, the Tasmanian, more closely related to the "highest" primate, the gorilla, than to the "highest" man, the European. Nonetheless, this was part of what Darwin described in the 1882 edition of *Descent of Man* as Haeckel's full discussion of "the genealogy of man." Indeed, Darwin praises Haeckel's work, writing that "almost all the conclusions at which I have arrived I find confirmed by this naturalist, whose knowledge on many points is much fuller than mine."[15]

Twelve years after the publication of *Origin*, Darwin, in *Descent of Man*, still could not answer the most important question raised by his own thesis: What is a species? Darwin's frustration over this definitional conundrum boils over in chapter seven, when he states:

> The question whether mankind consists of one or several species has of late years been much discussed by anthropologists.... But it is a hopeless endeavour to decide this point, until some definition of the term "species" is generally accepted; and the definition must not include an indeterminate element such as an act of creation.[16]

Darwin was here more categorical in dismissing creation as a factor in the definition of species than he had been in 1859. When he published *Origin* that year, he acknowledged that naturalists did not agree on a definition for species, "yet every naturalist knows vaguely what he means when he speaks of a species. Generally the term includes the *unknown element of a distinct act of creation*."[17]

Unable to define species, Darwin does no better with race. His concept of both terms, in fact, is muddled and arbitrary. He acknowledges the confusion between the two terms, stating that man "has given rise to many races, some of which differ so much from each other, that they have often been ranked by naturalists as distinct species."[18] In other words, naturalists are unable to define the difference between a species and a race.

For the evolutionist, attempts to classify according to species or race are arbitrary because living forms are continually undergoing evolutionary changes. This fluidity, Darwin held, makes it difficult to define the specific characteristics to use in defining a living form as a member of a species or race. Darwin noted that while naturalists who believe in evolution are convinced that all men "descended from a single primitive stock," they remain uncertain as to whether or not to "designate the races as distinct species, for the sake of expressing their amount of difference."[19]

Darwin, in fact, equivocated as to whether or not differences between people groups constitute different races or species. "It is almost a matter of indifference whether the so-called races of man are thus designated, or are ranked as species or sub-species, but the latter term appears the more appropriate."[20]

Darwin's discussion of race in chapter seven of *Descent of Man* is terribly confusing. He is unable to define the term race and is helpless to answer the basic question that he posed in the beginning of the book: What is "the value of the differences between the so-called races of man" and how did they originate?

Darwin failed to consider the possibility that there might not be any significant differences to measure between the "races." Beyond that, the method he used to approach the topic was totally wrong, since he was committed to viewing man "in the same spirit as a naturalist would any other animal."[21] Darwin's presumption that "common descent" was the means by which man—and all life—came to be, blinded him to the biblical account of the origin of the so-called races. The origin of the species, of course, is the "Creator," and the origin of "race" is recorded in Genesis 11.

We are all made in God's image and bear the mark of the benevolent Creator. Every creature, and especially man, is coded by the Creator with DNA, detailed chemical instructions for life that are inscribed into every cell. DNA is the reason living things reproduce after their kind (Genesis 1:12). DNA can produce a fig leaf, a roach, an elephant, and you and me. Humans give birth to humans, not ape-men. An elephant produces an elephant, not a rhinoceros; and, yes, a roach will produce a roach.

As a matter of fact, some female cockroaches mate only once and become pregnant for the rest of their lives. The bad news is that roaches never "evolved"—a fact that even evolutionists recognize. They say the cockroach came to be 280 million plus years ago and has never changed since. It still has eight hairy legs and 18 knees, holds its breath for 40 minutes under water, and runs three miles an hour—pretty fast for an insect. A roach will always be a roach because of DNA—God's code for life.

Swedish botanist Carl Linnaeus (1707-1778) has been justly called the "Father of Taxonomy" for his development of a binominal classification system that is still used today. He believed in the fixity of species and identified the biblical word "kind" for his definition of species. His system of classification was developed from a biblical perspective, as a means of identifying the different "kinds" made by the Creator. Had Darwin not rejected the Creator, he may have been more open-minded to seeing that all living things are actually marked by God with boundaries, and that if you traced creation back to its origin, you would find yourself at the Creator's feet.

The origin of the so-called races can be found in Genesis 11:9: "There the LORD confused the language of all the earth; and from there the LORD scattered them abroad over the face of all the earth." Distinct languages give rise to distinct cultures and to isolation from dissimilar groups. This isolation tends to cause people to marry in the same group and, in doing so, to develop physical characteristics unique to that group, such as skin and hair color, eye shape, facial features, and many others. The more the human genome is studied, the more evident it becomes that the effort to distinguish people groups according to what is loosely defined as race has no biological meaning.

The theory of evolution becomes extremely dangerous when applied to human beings. Any attempt to classify man within a system of living things puts the evolutionist in the position of determining the value of other men. It opens the door to ranking human beings as inferior or superior, depending on their place on the evolutionary scale. Superficial characteristics such as skin color, then, become clues to one's evolutionary progress and a justification for racial prejudice. Evolutionists are empowered to toy with absurd questions, such as whether Asians, Africans, Caucasians, and others can be classified as distinct species and, if so, whether they have equal value.

But if humans have differing values based on their evolutionary progress, what is to stop someone from eliminating races of humans that he thinks are lagging behind in the evolutionary contest? What is to stop some from purging the human gene pool to raise up a pure race, one that is more advanced in its evolution (like Haeckel's European man)? This is exactly what Adolf Hitler did. He applied Darwin's idea in Nazi Germany. He eliminated Jews, the handicapped, Gypsies, and anyone else the Nazis thought would get in the way of their coming "super race." Once the Creator is dismissed, mankind is placed in mortal danger.

No Genes in the Race

Darwin believed that race is a product of evolution and elaborated his view in *Descent of Man*. However, the idea of race—divisions within the human family based on characteristics such as skin color—is both unbiblical and has been set aside by genetics.

The Scripture teaches that we are all the sons and daughters of Adam. The only biblical distinctions to be made between members of the family of man are those that have to do with language and location. Paul told his audience of Greek intellectuals in Athens that

> He has made from *one blood* every nation of men to dwell on all the face of the earth, and has determined their preappointed times and the boundaries of their dwellings, so that they should seek the Lord, in the hope that they might grope for Him and find Him, though He is not far from each one of us (Acts 17:26-27 emphasis added).

Moreover, the work of Luigi Luca Cavalli-Sforza, an Italian geneticist who taught at Stanford University, "challenges the assumption that there are significant genetic differences between human races and, indeed, the idea that 'race' has any useful biological meaning at all," according to *The Economist*.[22]

An expert in human genetic variation who has committed his life to this study, Cavalli-Sforza has made remarkable findings which shed new light on the concept of race and undermine Darwin's idea that humanity can be divided into categories by race or sub-species.

Racial classifications are largely based upon superficial physical characteristics: eye and hair color, facial form, shape of nose, and skin color. These distinctions are only skin deep, says Cavalli-Sforza. "It is because they are external," he writes, "that these racial differences strike us so forcibly, and we automatically assume that differences of similar magnitude exist below the surface in the rest of our genetic makeup. This is simply not so; the remainder of our genetic makeup hardly differs at all."[23] Interestingly, Cavalli-Sforza found that there is more genetic variability between people within the same "racial" group than between different "racial" groups.

It was on the question of evolution and race that Darwin began to walk a tightrope. His "naturalism explicitly cast the notion of race into evolutionary and biological terms, reinforcing contemporary ideas of a racial hierarchy that replicated the ranking of animals."[24] Darwin detested slavery, but by his theory he breathed life into racist attitudes that were at war with the equality of all mankind. And Darwin himself accepted the notion that there is a top-to-bottom hierarchy among humans. Fortuitously enough, he was, as an Englishman, at the top of the evolutionary heap—Europeans being first on his evolutionary scale.

Darwin displayed his assumption of racial superiority in *Descent of Man* in a discussion about the evolutionary future of man. Below, he refers to the "civilized races" and the "savage races" of man and expresses the confidence that "civilized" men will, in the future, "certainly exterminate and replace the savage races." It is important to quickly acknowledge that Darwin was not here arguing for genocide. Still, not just his unfortunate phrase, but his theory itself, later animated those who were eager to do the work of evolution in more rapid and severe fashion.

> At some future period, not very distant as measured by centuries, the civilised races of man will almost certainly exterminate, and replace, the savage races throughout the world. At the same time the anthropomorphous apes, as Professor Schaaffhausen has remarked, will no doubt be exterminated. The break between man and his nearest allies will then be wider, for it will intervene between man in a more civilised state, as we may hope, even than the Caucasian, and some ape as low as a baboon, instead of as now between the negro or Australian and the gorilla.[25]

According to Darwin, man made it to the top of the tree of life after millions of years of struggle. But Darwin's elaborate arguments notwithstanding, his tree of life has no substance to prove its existence. There is no scientific evidence to support it. The common descent of man is a fabrication. Its exponents are blind to the fact that similarity in design is the consequence not of common descent, but a common Creator.

Darwin and his disciples denied the evidence for the Creator and erected an idol in the shape of the tree of life. Those who followed Darwin revered him as they embraced, popularized, and applied his ideas. These disciples of Darwin applied his ideas in Britain, Germany, and America. Thomas Huxley was an effective popularizer of Darwinian evolution in Victorian England. Ernst Haeckel brought the idea of evolution into Germany—an idea that eventually found its way to Nazi death camps. Francis Galton, a cousin of Darwin, came up with the idea of eugenics, which became popular in America. Darwin's tree of life was a dangerous idea. It was an idea that would kill millions.

CHAPTER SIX

Darwin's Disciples

Three leading disciples of Darwin—Thomas Huxley, Ernst Haeckel, and Francis Galton—popularized and advanced the idea of evolution. These three contemporaries proved to be faithful acolytes who both proselytized for and applied Darwin's theory in new and disturbing ways.

It is not surprising that it fell to others to promote the Darwinian revolution. Darwin had no appetite for confrontation. In fact, it took some persuading from his friends to convince him to even publish his theory. Although they reassured him that all would be fine if he did so, he had doubts. He knew that his theory challenged the authority of God and the Bible and had every expectation that he would be subjected to ridicule or "execrated as an atheist."[1]

Darwin's publisher, John Murray, was himself not sure how the book would be received. Darwin thought the audience for his book would be limited to the professional scientist and amateurs with an interest in scientific matters. Murray conservatively printed 1,250 copies, which sold out the first day. Another 3,000 copies were printed, and there were six more editions that followed—the last one published in 1876. At Darwin's death in 1882, a total of 24,000 copies had been sold.[2]

Origin was in demand internationally as well. French, German, Italian, Russian, Spanish, Polish, Bohemian, and Japanese translations were produced in short order. There was a hunger for Darwin's theory of evolution because, particularly in Europe, there was a need to fill the vacuum once occupied by God. Darwin's tree of life offered to fill the void created by the abandonment of God.

Darwin was a country gentleman who was extremely careful in how he faced the public. Indeed, as he developed his theory, after returning from the *Beagle* voyage, he was "feeling jumpy about the hysteria his views would unleash among his clerical friends."[3] Some 20 years later, when he published *Origin*, Darwin sent advance copies to friends and professional colleagues with notes that underscore his anxiety and his recognition of the impact his work would have. To Henslow, his friend and tutor at Cambridge, he wrote, "I fear you will not approve of your pupil in this case." To Oxford geologist John Phillips, who had succeeded William Buckland at Oxford, he wrote, "You will 'fulminate anathemas.'"[4]

At ease in his study, but not in public campaigning for his theory, Darwin needed the aid of those better equipped to do the work of evangelizing on behalf of the evolutionary tree of life. While he had dedicated his life to the study of evolution, he was not a great orator, did not enjoy debate, and besides all this, was in continual poor health.

Darwin's "Bulldog": Thomas Huxley

The publication of *Origin* in 1859 began to produce the much-anticipated following that Darwin needed to spread the "truth" of evolution and do away with God. Many Darwin enthusiasts emerged in support of natural selection. One man stood out for his charisma and a seemingly inexhaustible store of energy that would be used to persuade countless people to convert to evolution. His name was Thomas Huxley, Charles Darwin's "Bulldog."

Although they had corresponded on professional concerns, Thomas Huxley and Charles Darwin first met at The Geological Society in April 1853. Both men were recognized by the prestigious Royal Society for their accomplishments: Huxley, in 1852, for his work on Hydrozoa (creatures in the ocean, including Portuguese Man-o-War), and Charles Darwin for his work on barnacles in 1853.

Huxley was extremely bright. He passed his medical boards at age 20 and won a gold medal for work in anatomy and physiology. He was a highly curious man who studied marine life while serving in the British Admiralty as a surgeon. After he left the British navy, he became friends with Herbert Spencer who, like Huxley, was a "freethinker." Spencer, who coined the term, "Social Darwinism," became a well-known philosopher and socialist who applied evolutionary theory to physiology, philosophy, and sociology.

Huxley became a lecturer and naturalist at the Royal School of Mines in 1854 and served in that capacity for 31 years. His lectures were well attended and his charismatic style kept everyone attentive, including lay people, who eagerly listened. Darwin thought highly of Huxley. He wrote in his *Autobiography*:

> His mind is as quick as a flash of lightening [sic] and as sharp as a razor. He is the best talker whom I have known. He never writes and never says anything flat. From his conversation no one would suppose that he could cut up his oppo-

> nents in so trenchant a manner as he can do and does do.[5]

Darwin had what Huxley needed, a mechanism for evolution, and with this, the two men formed a cohesive unit to powerfully influence Britain and Western civilization.

Huxley was effusive after he read *Origin*, thanking Darwin the day after the book went on sale "for the great store of new views you have given me...." Huxley pronounced himself "prepared to go to the Stake if requisite" to defend Darwin's theory and announced that he was ready to answer the "considerable abuse & misrepresentation which, unless I greatly mistake, is in store for you...." Eager for the contest to begin, Huxley let Darwin know, "I am sharpening up my claws and beak in readiness."[6]

Huxley had the opportunity to do so in June 1860, when he clashed with Samuel Wilberforce, Bishop of Oxford, in a famous debate at Oxford. This encounter, which Darwin missed due to ill health, has been cast in its retelling as a great victory for Huxley and Darwinism in the contest between science and religion. While initial accounts of the debate do not bear out the legend that Huxley dispatched "Soapy Sam," delivering a cool, reasonable, and overwhelming response to Bishop Wilberforce's casual joke about Huxley's primate origin, that has become the received account of the occasion.

British botanist Joseph Hooker, also a debate participant, may have more notably bested Wilberforce in the debate than Huxley, according to his own after-action report in a letter to Darwin. Nonetheless, Hooker spoke highly of Huxley's contribution years later:

> The famous Oxford Meeting of 1860 was of no small importance in Huxley's career. It was not merely that he helped to save a great cause from being stifled under misrepresentation and ridicule—that he helped to extort for it a fair hearing; it was now that he first made himself known in popular estimation as a dangerous adversary in debate—a personal force in the world of science which could not be neglected. From this moment he entered the front fighting line in the most exposed quarter of the field.[7]

The debate, with Wilberforce's attempted jest about Huxley being the progeny of primates, brought into focus the question of man's place in the evolutionary program—something Darwin did not address in *Origin*. Eleven years later, Darwin offered his readers his conclusions on the question of ape-man evolution when he published *Descent of Man*.

Huxley wanted to address this subject more rapidly.

Five years after *Origin*, he published his *Evidence as to Man's Place in Nature*, the first attempt to chart man's alleged evolution. The book examined primate and human paleontology and presented what he considered to be evidence for human evolution. In it, Huxley took on anatomist Richard Owen, an opponent of Darwinism, who asserted that the brains of apes and men were distinct.

Perhaps the most impressive feature of Huxley's book was a line drawing he had commissioned. This illustration of five skeletons purported to show the course of evolution from gibbon to orangutan, chimpanzee, gorilla, and man. This "inspired visual propaganda," as Darwin biographer Janet Browne calls it, relied on skeleton figures that were not all to the same scale, in order to more effectively demonstrate an alleged progression from monkey to man.[8] However inaccurate the illustration, the drawings have been effectively used to show evolution. Darwin welcomed this work of his "Bulldog," stating, "Hurrah, the Monkey Book has come" and expressed a keen interest in reading it.

The application of evolution to man led in Huxley's thought to the assumption of "inferior" races of men. Huxley took it as an undoubted verity that the black man was inferior to the white man. He stated, for example, that:

> No rational man, cognizant of the facts, believes that the average negro is the equal, still less the

> superior, of the white man. And if this be true, *it is simply incredible* that, when all his disabilities are removed, and our prognathous [having a projecting lower jaw] relative has a fair field and no favour, as well as no oppressor, he will be able to compete successfully with his bigger-brained and smaller-jawed rival, in a contest which is to be carried out by thoughts and not by bites.[9]

Huxley came out not only for the theory of evolution, but also against the idea of the soul and miracles. He invented the term "agnostic" to describe his attitude toward God and Christianity. This view, which claims uncertainty about the existence of God and allows inquiries into that question, was popularized by Huxley.

Huxley's aggressive opposition to faith is illustrated by his Sabbath-day series of lectures on evolution, to which he attracted large crowds. These "Sunday evenings for the People" lectures were so popular that in January 1866, on the first Sunday, 2,000 people were turned away.[10] Darwin's disciple, Thomas Henry Huxley, offered escape from "fine-spun ecclesiastical cobwebs"[11] as he proclaimed a form of material salvation.

The most effective, aggressive, and influential of Darwin's defenders, Huxley was joined by others who advanced and applied Darwin's ideas in new ways not anticipated, perhaps, by Darwin himself.

DARWIN'S COUSIN: FRANCIS GALTON

It changed his life. Francis Galton was 37 years old in 1860 when he read *Origin of Species*. The book, Galton wrote, "made a marked epoch in my own mental development, as it did in that of human thought generally." Most notably, Darwin's book liberated him from the authority of the Church. "Its effect was to demolish a multitude of dogmatic barriers by a single stroke," Galton wrote, "and to arouse a spirit of rebellion against all ancient authorities whose positive and unauthenticated statements were contradicted by modern science."[12]

Thirteen years younger than Darwin, Francis was Charles' first cousin and a worldwide traveler and discoverer. Like his cousin who traveled on the *Beagle* to search out the unknown, Galton was motivated by a passion to venture where no other man had gone. Galton's father, a prosperous banker, had left a sizeable inheritance to his son who, without hesitation, used the money to follow his dreams and travel the world.

He took traveling seriously. Galton consulted with the Royal Geographical Society and ambitiously followed the path of the famous English missionary explorer David Livingstone, tracking into the deep jungles of Africa. He led an expedition to find a passage to the newly discovered Ngami Lake in the heart of the African jungles.

Although he did not accomplish his mission, he brought back valuable information and was accorded respect on his return as an explorer and geographer. A bright, ambitious young man, Galton published in 1853 his *Narrative of an Explorer in Tropical South Africa.* A thorough work, it made new contributions to charting Africa, and helped win his election to the prestigious Royal Geographical Society.

Francis Galton was born in 1822 into a Quaker home. His mother was a daughter of Erasmus Darwin, the highly intelligent, radical anti-God evolutionist, and the stepsister of Robert Darwin, Charles' father. Galton's father was a banker from a successful and wealthy family of bankers and gunsmiths. Francis at first studied medicine at Trinity College at Cambridge, but moved on to major in mathematics because he enjoyed a more rigorous analytical discipline to being involved in patient care.

Galton had the capacity to be both a creative thinker and an analytical problem solver. He applied his gifts to the study of weather and made significant contributions to meteorology. He demonstrated methods for mapping weather and coined the term "anticyclone," which is still used today. A habitual measurer for purposes of prediction, Galton identified human fingerprints and created a system for identifying the sworls unique to each individual. His book, *Finger Prints*, described the new discipline of fingerprinting and featured Galton's own fingerprints on the title page.

Already successful, well established, recognized for his contributions, and fully engaged in his many endeavors, Galton underwent a complete transformation when he read his cousin's book. He determined to follow in Darwin's footsteps and dramatically reordered his life to focus on the implications of evolution. His admiration for Darwin knew no bounds. After Darwin died, he wrote, "There was no man who I reverenced, or to whom I owed more, *spiritually*, than to him. His *Origin of Species* first put me, so to speak, in harmony with Nature."[13]

While Huxley was Darwin's public defender, Galton focused on how best to advance evolution, using sexual selection as a means to guide the process of evolution in civilized society. He devoted himself to the study and measurement of human characteristic, even mapping the location of beautiful women in Britain. "Galton distinguished himself by his ability to recognize patterns, making him an almost unique connoisseur of nature," according to Edwin Black, author of *War Against the Weak*.[14]

He outlined his ideas in an article for *Macmillan Magazine* in 1865 and published his more full-blown theory four years later in his *Hereditary Genius*.[15] In this work, Galton made the case that ability is passed down by human descent. To offer proof, he chronicled the heredity of many of England's most distinguished judges, soldiers, statesmen, scientists, and literary men.

"Judicious Marriages"

If, as Galton claimed to show, ability is derived by inheritance, then it becomes possible to breed men just as one breeds dogs or horses. Galton explained:

> Consequently, as it is easy ... to obtain by careful selection a permanent breed of dogs or horses gifted with peculiar powers of running, or of doing anything else, so it would be quite practicable to produce a highly-gifted race of men by judicious marriages during several consecutive generations.[16]

Darwin congratulated Galton on this work, telling his younger cousin in a letter that, "I do not think I ever in all my life read anything more interesting and original."[17] Nonetheless, contemporary apologists of Darwin seek to distance Darwin from those who made social application of his ideas. Niles Eldredge thinks that "Darwin would cringe at some of the movements undertaken in his name." Eldredge does, however, acknowledge that "social Darwinism," which he regards as an illegitimate offspring of Darwin's theory, "has given us the eugenics movement and some of its darker outgrowths, such as the genocidal practices of the Nazis in World War II—where eugenics was invoked as a scientific rationale to go along with whatever other 'reasons' Hitler and his fellow Nazis had for the Holocaust."[18]

Galton, who displayed brilliance in travel, meteorology, statistics, and in the use of fingerprints for identification, is the founder of the eugenics movement. He coined the term "eugenics" (well born) in 1883, and campaigned extensively for "judicious marriages," which, he thought, would, over several generations, produce a highly gifted race of people. He defined eugenics as "the science which deals with all influences that improve the inborn qualities of a race; also with those that develop them to the utmost advantage."[19]

Galton believed that heredity is the major biological determinant in the "upward movement" of evolution. But heredity, he said, required some measure of social engineering to augment the "favored stock." Galton wrote that, "The possibility of improving the race of a nation depends on the power of increasing the productivity of the best stock."[20] In other words, it is possible to breed humans to make a better world. Galton wanted to limit marriage to the union of well-born partners and prohibit marriages of the "unfit." By so doing, he thought that "what Nature does blindly, slowly, and ruthlessly, man may do providently, quickly, and kindly."[21]

Galton's eugenics were imported into America by biologist Charles Davenport, with the financial support of the Carnegie Institute. Davenport served as a director of the American Eugenics Society, which had on its board several individuals who also served on boards of organizations linked to Margaret Sanger. Sanger, the founder of Planned Parenthood, championed so-called "birth control" for eugenic purposes. According to Sanger, "Birth Control has been accepted by the most clear thinking and far seeing of the Eugenists themselves as the most constructive and necessary of the means to racial health."[22]

The eugenics movement held that intelligence and good education is a source of morality. That proposition served as a basis on which Lewis Terman and other eugenicists developed the IQ test. This test assigned the terms "morons," "imbeciles," and "idiots" to those who scored less than 70 on the IQ scale. "If we could preserve our state for a class of people *worthy* to possess it, we must prevent, as far as possible, the propagation of mental degenerates,"[23] said Terman. He believed, as did Galton, that society should identify and eliminate the weak and the feebleminded. The IQ test was designed to do that.

Francis Galton replaced God with Darwinian evolution and declared that there was no need for religion because science is a valid alternative.[24] In so doing, this disciple of Darwin devalued man and helped to weaken a moral brake that could have stopped the Nazi atrocities that followed. In his essay, "Eugenics as a Factor in Religion," Galton laid out arguments that would lead to Nazi killings in the years to come. And he left no doubt about the link between evolution and eugenics:

> The creed of eugenics is founded upon the idea of evolution; not on a passive form of it, but on one that can to some extent direct its own course.... Evolution is in any case a grand phantasmagoria, but it assumes an infinitely more interesting aspect under the knowledge that the intelligent action of the human will is, in some small measure, capable of guiding its course. Man could do this largely so far as the evolution of humanity is concerned....[25]

Eugenics and evolution cannot be separated. One leads logically to the other. Galton believed in active intervention because man is the most intelligent part of the tree of life and therefore should direct his own destiny.

The German Darwin: Ernst Haeckel

Darwin's *Origin* was classified as a science book, but it had broad appeal. It was not just found in the hands of the scientific elite and academicians, but attracted public attention. *The Saturday Review* wrote that the book found its way "into the drawing-room and the public street."[26] Darwin was amazed that his book sold out the first day and that people from all walks of life became enamored with his work.

It was not just popular, but also extremely controversial. It was a direct attack against the Church and the authority and reliability of the Bible. Darwin's Bulldog, Thomas Huxley, explicitly framed the debate in terms of a conflict between science and the Bible. Huxley wrote:

> Extinguished theologians lie about the cradle of every science as the strangled snakes beside that of Hercules; and history records that wherever science and orthodoxy have been fairly opposed, the latter have been forced to retire from the lists, bleeding and crushed if not annihilated; scotched if not slain.[27]

For Huxley, evolution was science. The claim that evolution was a scientific idea was a way for Darwin and Huxley to give the new theory much-needed credibility and greater acceptance. In fact, science and evolution are distinct. Unlike true science, the claims of Darwinian evolution cannot be tested or replicated. That is because origin events happened just once and are, therefore, outside the purview of scientific experiments.

Darwin's controversial new idea shot across Europe like a bolt of lightning. Ernst Haeckel, a German physician with a small practice, was profoundly moved and changed after reading a German translation of Darwin's *Origin*. Later in life, he wrote his mistress that he became a freethinker and pantheist after he studied Darwin's book.[28]

Haeckel, who was 25 years younger than Darwin, fairly worshipped the British naturalist and the idea of evolution that he brought to the world. Since Haeckel's interest in medicine paled before his love for natural science, *Origin* proved a powerful impetus to redirect the path of his life. He quickly emerged as the chief apostle of Darwinism in Germany. Haeckel is recognized for being "the first German biologist to give a wholehearted adherence to the doctrine of organic evolution and to treat it as the cardinal conception of biology."[29]

Haeckel's academic career began in 1861 at Jena University, where his lectures on Darwinism attracted classes of 150 or more. He was an ambitious, bright young man, appointed at age 31 to the prestigious position of chair of Biology, a post he held for 43 years despite many other lucrative offers.

Darwin began corresponding with Haeckel in 1865, and acknowledged the great influence he was having in Germany. Haeckel paid a visit in 1866 to Darwin at his home, which to the 32-year-old devotee was like a religious experience. It was difficult to understand Haeckel's broken English, but Darwin was impressed and stated that he "had seldom seen a more pleasant, cordial and frank man."[30] Haeckel visited Darwin on several other occasions, visits which, Darwin's son wrote, "were thoroughly enjoyed by my father."[31] Darwin was extremely encouraged by Haeckel's work in Germany and the acceptance there of his theory. The progress of his ideas in Germany, he wrote, was the "chief ground for hoping that our views will ultimately prevail."[32]

Darwin biographers Adrian Desmond and James Moore describe Ernst Haeckel as "bombastic."[33] He enthusiastically made a broad social application of Darwin's theory, something easy to do because Darwinism is a philosophy despite its proponents' claim that it is science. Haeckel, in fact, zealously applied evolution and moved into arenas "that Darwin feared to tread in."[34] Haeckel openly opposed biblical truth. He castigated the biblical idea of the immutability of species as a "colossal dogma…empowered by blind belief in authority" and spoke of the Creator-God as a "gaseous vertebrate."[35] While Darwin himself had parted ways with Christianity, he wanted no part in Haeckel's open, strident attack against Christian faith.

Haeckel: The Good, the Mostly Bad, and the Hideously Ugly

The legacy of Ernst Haeckel can be divided into the good, the mostly bad, and the hideously ugly. The "good" is that, using the microscope, he was able to create illustrations that showed, in exceedingly fine detail, marine life that had never been seen before. Haeckel, an accomplished artist, used his artistic ability to create his *Art Forms in Nature*, a printed collection of his nature illustrations. He also coined the terms phylum, phylogeny, and ecology, all of which are used by biologists today.

The "mostly bad" defines his work to prove evolution, which was proved false or fraudulent by the end of his life, or shortly after his death. Haeckel exaggerated the data, doctored images, censored information, used only what would support his conclusion, and created deceptive drawings that were complete misrepresentations.

Science writer Eric Weisstein, among others, has pointed out that Haeckel's book, *Natürliche Schöpfungsgeschichte (The Natural History of Creation)*, was shown to be fraudulent in 1868, the same year it was published. Nonetheless, science textbooks continued to discuss Haeckel's views as legitimate science into the 1990s.

Haeckel's *Natural History of Creation* expanded on Darwin's theory. It showed illustrations of embryos of various invertebrates in an attempt to document Haeckel's bizarre theory that early embryos are physically alike and that their developmental stages resemble the course of their own evolution.

To make his case Haeckel showed woodcut prints, which he said were of dog and human embryos. Nine pages later, Haeckel placed prints which he said were of tortoise, chicken, and dog embryos and claimed they looked alike. They did, but that was because the drawings were all from the same woodcuts.

This fraud was exposed in 1868 by L. Rutimeyer, a zoology and comparative anatomy professor at the University of Basel. According to Rutimeyer, "There is considerable manufacturing of scientific evidence perpetrated. Yet the author has been very careful not to let the reader become aware of this state of affairs."[36]

Wilhelm His, a well-known embryologist and professor of anatomy at the University of Leipzig, also uncovered the fraud. He showed in 1874 that Haeckel had doctored the drawings presented in his book. He was scathing in his judgment of Haeckel, stating that Haeckel had disqualified himself as a science researcher by his blatant fraud.[37]

With Haeckel exposed as a fraud in 1868, it is amazing that Darwin cited Haeckel as a resource in his 1871 book, *The Descent of Man*. He wrote there of Haeckel that, "almost all the conclusions at which I have arrived I find confirmed by this naturalist [Haeckel], whose knowledge on many points is much fuller than mine."[38] Darwin also praised Haeckel in an 1867 letter to the German scientist, written after he read his *General Morphology*:

> It never occurred to me for a moment to doubt that your work, with the whole subject so admirably and clearly arranged, as well as

> fortified by so many new facts and arguments, would not advance our common object in the highest degree."[39]

In a November 1868 letter, Darwin complimented Haeckel for his work in *Natural History of Creation* in creating a very detailed genealogical tree of life. At the same time, he expressed uneasiness about his scientific audacity. He wrote:

> Your chapters on the affinities and genealogy of the animal kingdom strike me as admirable and full of original thought. Your boldness, however sometimes makes me tremble, but as Huxley remarked, someone must be bold enough to make a beginning in drawing up the tables of descent.[40]

Haeckel's *Natural History of Creation* contained enormous speculation about the tree of life that went beyond Darwin's *Origin*. Even so, Huxley approved, stating that any effort to advance Darwin's theory, even if in the wrong direction, was preferable to no movement at all. Huxley wrote:

> In Professor Haeckel's speculations on Phylogeny, or the genealogy of animal forms, there is much

> that is profoundly interesting, and his suggestions are *always supported* by sound knowledge and great ingenuity. Whether one agrees or disagrees with him, one feels that he has forced the mind into lines of thought in which it is more profitable to go wrong than to stand still.[41]

Huxley himself ultimately adopted Haeckel's method and created his own version of the tree of life. His "genetic classification," which went back to the dinosaurs, was an attempt to show how "all living beings have been evolved one from the other."[42]

Haeckel was prolific. By age sixty he had written 42 works containing 13,000 pages, most of which were highly dubious, if not fraudulent. The foundation of Haeckel's tree of life was a simple protoplasm he claimed to have identified, which he called *Monera*. This basic unit of life, he claimed, arose from carbon and nitrous carbon (carbon with nitrogen) in its complex form and moved from inorganic matter to organic matter by spontaneous generation. He produced elaborate drawings of *Monera*, which, as he knew, was non-existent. He had the audacity to publish 73 pages of his speculations, with 30 fake drawings of these *Monera*, along with their scientific names, in a German scientific journal.[43]

Haeckel's imagination was placed fully in the service

of Darwin. He drew evolutionary trees, falsified embryos, created racist pictures of men and apes, and he did it all from his imagination—yet with the claim of scientific authority. He went so far as to propose the existence of a "missing link"—an ape-man creature he called *Pithecanthropus alalus*, or silent ape. He had a drawing produced of a male and female *Pithecanthropus* in domestic repose. It was all imaginary, fictional, and fraudulent. Haeckel's primary theoretical innovation, his so-called biogenetic law, which asserts that embryonic development repeats the stages of evolution, has been discredited for more than a century.

The "hideously ugly" aspect of Haeckel's legacy lies in the manner in which he applied his evolutionary concepts to human ethics and the value of human life. Evolution denies the existence of the Creator. When the Creator is dismissed, so are the absolute rules laid down by a rational and all-intelligent Creator. In that event, the world becomes a much more dangerous place in which to live.

Haeckel, Darwin, Huxley and all those who accept evolution for what it is, have to submit to a moral and ethical system in flux. If there are no fixed, uniform, and universal moral principles that apply to all men at all times, a culture will quickly lose its moral anchor. If there are no God-given moral principles to govern mankind, the fragile, man-made moral system that replaces that given by

God is one in which every evil becomes possible.

It is important to understand that Darwinian evolution not only has no scientific credibility, but it is the product of human rebellion against the benevolent Creator. It is simply an anti-God philosophy dressed up as science. Haeckel denied the existence of a personal God and attacked the Creator God as a "gaseous vertebrate."[44] He was a popular, influential German scientist, the Darwinian apostle of Europe who, cloaked with scientific respectability, used pure fraud to promote his anti-God agenda.

Haeckel did not let the facts of science lead him to the truth. Instead, he twisted science to follow his ideology. Europe, including Germany, was fertile ground for the atheistic Darwinian revolution that sought to dethrone God. That receptivity made it vulnerable to evil unleashed.

Many of Haeckel's books and publications were written with artistic flare for the common man. They were highly popular and widely read in Germany. His books, of course, were filled with fallacious material, but readers devoured his work, and wanted more.

Haeckel's moral system centered on the evolutionary belief that morality changes with the times. He believed that humanity is the product of a process of development predetermined by biology. He held that every aspect of a human being, including his soul—his thoughts, will, and emotions—is just the product of fortuitous protoplasmic change over time. This view had, not too surprisingly, a racist dimension to it. Haeckel believed that the different "races" are really diverse species and display varying stages of evolutionary development. Therefore, humans have varying intrinsic worth depending on how evolved they are. Humans of higher worth demonstrate their more evolved state by how well they produce significant cultural achievements because of their inherited biological traits, such as intelligence, diligence, and moral character.

Haeckel believed that humans were not all equal. In fact, some humans were more closely linked to the most evolved animal than they were to the most evolved human. As Haeckel put it, "the differences between the lowest humans and the highest apes are smaller than the differences between the lowest and the highest humans."[45]

Haeckel likened Australian aborigines and the Bushmen of South Africa to the apes. This led to some grotesque attempts to crossbreed. A gentleman influenced by Haeckel's ideas consulted Haeckel about an attempt to use artificial insemination to cross an ape with a "Negro." Approached by a German sexologist about another similar project, Haeckel recommended "trying to inseminate a chimpanzee with the sperm of a black African."[46]

Haeckel's "hideously ugly" contribution to human affairs is even more pronounced in efforts to purify the race by getting rid of the weak. He favored infanticide for babies with congenital deformities, disabilities, or mental retardation. Babies, he thought, were in a primeval evolutionary stage and could be easily disposed of because they were similar to unwanted animals. He suggested "a small dose of morphine or cyanide"[47] would free parents from the burden of taking care of these children for the rest of their lives. Why waste energy on those who will not contribute to society?

While Haeckel believed life begins at conception, the unborn human did not, in his view, acquire the worth and status of a human until some point after birth. Abortion and infanticide were equivalent, he believed, to killing beings at lower stages of evolutionary development. Haeckel wrote that, "the developing embryo, just as the newborn child, is completely devoid of consciousness, is a pure 'reflex machine,' just like a lower vertebrate."[48] Haeckel's attitude is shared by contemporary advocates of abortion, who describe the unborn child as a "product of conception."

Haeckel's doctrine of eliminating the weak applied equally to the mentally ill. Since these unfortunates never reach the higher consciousness and intellect of most humans, Haeckel believed they should be regarded as of lesser worth and their lives ended. He believed in forced euthanasia. Human life that is worthless and does not benefit the race should be done away with, in his view. Those he regarded as leading "worthless" lives included lepers, cancer patients, and those with other incurable diseases.[49]

When it came to ranking the different "races," Haeckel considered Europeans to be the superior race. Writing in 1917, he took extreme offense that the enemies of Germany were using non-European troops against them. Haeckel called this an "underhanded betrayal of the white race."[50] According to Haeckel:

> A single well-educated German warrior, though unfortunately they are now falling in droves, has a higher intellectual and moral value of life than hundreds of the raw primitive peoples, which England and France, Russia and Italy set against us.[51]

Haeckel endorsed German colonialism and annexation of other European territories. The struggle to survive included the taking of lands to preserve the favored race. He also believed in the extermination of the "inferior" races. Haeckel's pseudo-scientific justification for German military aggression, as well as eliminating the sick, weak, and unfit, gained a broad audience in Germany and helped lay the foundation for Hitler's Nazi ideal. The result was the death of millions of innocent lives, swept away by an evil idea rooted in Darwin's theory of evolution. The outcome was hideously ugly.

CHAPTER SEVEN

THE PATH FROM DARWIN TO HITLER

Just seventeen days after *Origin* went on sale in Britain in 1859, German academic Heinrich Bronn was already planning a German translation. Bronn, a prominent German geologist, published a freely translated edition of *Origin*, to which he added his own sometimes critical commentary on Darwin's ideas. Bronn used his German version of Darwin's *Origin* to show that evolution is necessarily linked to spontaneous generation, the idea that organic matter can arise from inorganic matter. The translation was issued at a time when a scientific controversy continued in Paris over the question of spontaneous generation, despite the fact that Louis Pasteur had shown by means of a series of experiments that life could not arise from non-life.

Bronn also added a chapter, identifying the religious problems created by Darwin's theory. He said that unless it can be proved that life can come from non-life, "readers must consider descent with modification an unproven suggestion."[1] Darwin was not pleased with this free translation, which was published in 1860 to mixed reviews. It was followed in 1867 by a more faithful, word-for-word translation by a strong Darwinian evolutionist, Professor Julius Victor Carus, who worked closely with Darwin in translating the fourth edition.

Origin entered German culture at a time when the anti-Christian ideas of Ludwig Feuerbach were already widely disseminated among Germans. Feuerbach, who helped prepare the German mind for evolution, had published *The Essence of Christianity* in 1841. This skeptical treatise, which went rapidly through three editions, declared that "the idea of God was a dream," a construct of human imagination and the product of "human longing for security." Feuerbach argued that Christianity was a "false divine mandate for their personal advancement."[2]

While Feuerbach's atheism did not win broad public support, his views held sway with an influential minority in higher education and among the wealthy. When *Origin* arrived in Germany, a powerful audience was in place to enthusiastically welcome this "scientific" argument for atheism. Darwin's work proved more compelling than Feuerbach's because the German skeptic used wooly abstract arguments, while Darwin offered concrete observation from nature. His arguments arose from his observations—and his conjectures—about real natural phenomena.

Origin, of course, energized Germany's leading Darwinist, Ernst Haeckel. A prolific writer and artist, he became a renowned German author and public figure in late nineteenth and early twentieth century German society. His books *Natural History of Creation* (1868), *The Riddle of the Universe* (1899), and *The Wonders of Life* (1904) were German bestsellers that transmitted Darwin's ideas to both the intelligentsia and the German working class. Haeckel's Darwinian formulations were not only topics of discussion in academic circles, but were also batted about in German beer halls. His legacy gave impetus to Hitler and his followers, who found his Darwinian-based arguments for racism and nationalism persuasive. Haeckel's famous quote, "politics is applied biology," was widely used by Nazi propagandists and by Hitler himself.

Haeckel believed that the so-called lower races of humanity would be exterminated by natural selection. Europeans stood supreme and other inferior races, the

American Indian or the Australian aborigine, were doomed to extinction. He explained that, "Even if these races were to propagate more abundantly than the white Europeans, yet sooner or later they would succumb to the latter in the struggle for existence."[3] Haeckel, who favored German colonialism, wrote in his 1917 book, *Eternity*, that non-European troops were members of the "wild races."[4]

Richard Weikart, author of *From Darwin to Hitler*, a seminal work that traces in careful detail the links between the naturalist and the mass killer, notes that while Haeckel was the most prominent Darwinist to advocate racial extermination, he was not the only one to do so. Less than one year after *Origin* was published, the editor of a German journal on geography and ethnology embraced Darwinism and the idea of race competition and the extinction of the inferior races. Oscar Peschel, editor of *Das Ausland*, said Spaniards were not to blame for killing Indians, since they were only acting in accordance with the law of natural selection. According to Peschel:

> This is the historical course. If we view it with the eye of a geologist, and indeed a geologist which accepts the Darwinian theory, we must say that this extinction [of human races] is a natural process, like the extinction of secondary animal and plant forms.[5]

Weikart notes that other writers in Germany besides Haeckel came under the spell of Darwin and offered their own twist on how the ideas in *Origin* offered freedom from the traditional restraints of the Church—and, in fact, liberation from God. Bartholomaeus von Carneri, a good friend of Haeckel's, was an Austrian politician who wrote *Morality and Darwinism* (1871) and other books which examined ethics and morality in the light of evolution. For Carneri, "the value of Darwin is that humans no longer need to have a supernatural soul, and that one no longer needs purpose to explain creation."[6]

The economist Albert E. F. Schäffle was, after Carneri, the most influential writer to apply Darwin to ethics in the 1870s. He held that morality is not God-given, but the outcome of the evolutionary process. "Law and morals necessarily arise in and through the selective struggle for existence," Schäffle wrote, "since they themselves are essential components of the power of collective self-preservation."[7]

Max Nordau, whose 1883 book, *The Conventional Lies of Our Civilization*, sold more than 50,000 copies by 1903, also believed that moral principles were the product not of transcendent revelation but of natural selection. The struggle for existence, he wrote, is the force that "shapes in its widest sense all of human history."[8] And if evolution, an ongoing process, shapes morality, then morality itself must be subject to change. Standards for right and wrong, he wrote, are "not absolute, but variable" and "are subject to the laws of evolution in society and therefore in a constant state of flux."[9]

Weikart also notes the influence of Italian psychiatrist Cesare Lombrosa, who believed that criminal behavior was hereditary. While Lombrosa's views were controversial in Germany, the idea, Weikart states, "of the 'born criminal,' or at least that heredity plays an important role in moral and immoral behavior, steadily gained ground in late nineteenth and early twentieth-century German psychiatry."[10]

The Darwinian ferment in Germany was such, according to Weikart, that by the "1890s and early 1900s Darwinism had become well-entrenched in Germany." By this time, Weikart writes, "Racial theorizing, most of which was laced with Darwinian rhetoric, was heating up, capturing the imagination of ever wider audiences."[11]

Hitler and Darwin

It was into this intellectual atmosphere that Adolf Hitler was born on April 20, 1889, seven years after the death of Darwin. Hitler was born in a small Austrian border town to Alois and Klara Hitler. Alois, an older man who married his maid, Klara, in 1885, was a customs agent who earned a modest middle-class income. He had an explosive temper and a history of womanizing and drunkenness.

Sickly at birth, Adolf held a special place in Klara's

heart because she had lost three children before his birth. Alois, who died in 1903 when Adolf was approaching fourteen, was overbearing and insisted that Adolf maintain his grades so that he could become a civil servant and follow in his steps. Adolf rebelled and did poorly in school. Biographer Robert Payne writes that Hitler "was a proud, explosive, and difficult youth, full of strange ideas and wild ways."[12]

Hitler left home in 1907 to study art in Vienna at the General School of Painting at the Academy of Fine Arts. He failed the school's entrance exam and had to return home to be with his mother, who died from breast cancer in December 1907. Hitler's deep love for his mother became evident at her death. The physician who cared for Klara wrote, "In all my career I have never seen anyone so prostrate with grief as Adolf Hitler."[13]

Hitler returned to Vienna to retake the art school entrance exam, only to fail again. He spent his small inheritance and began to drift aimlessly on the streets. Desperately poor, he spent his nights in a homeless asylum. He attempted to use his limited artistic talent to make postcard size paintings for sale, but his lack of self-discipline, mood swings, and inexplicable rages stymied his success.

Depressed and indigent, Hitler began to consume literature dealing with the occult, telepathy, phrenology, graphology, and other pseudoscientific topics. Most historians believe that it was at this time that Hitler became an admirer of Lanz von Liebenfels, editor of the magazine *Ostara.* Liebenfels propounded ideas about the favored blue-eyed, blonde Aryans who, he thought, were threatened by the "Untermensch," the lower, less-evolved races.

Liebenfels believed that Aryans were at the top of the race hierarchy with the black man and Mongolians at the bottom. These lower races were, he thought, to be the slaves of the Aryans. He spoke of the "Holy Grail of the German blood," which had to be protected by the racially pure. Hitler may have remembered that phrase because he used it in the 1930s when he told a fellow Nazi:

> The problem is this: how can we arrest racial decay?... Shall we form ... a select company of the really initiated? An Order, the brotherhood of Templars around the holy grail of pure blood?[14]

Liebenfels was also virulently anti-Semitic. In a 1908 pamphlet he spoke of Jews and others as a "mongrelized breed" and urged their elimination.[15]

Another Viennese race theorist, Guido von List, may also have impacted Hitler. List believed that Aryans were superior because their blood was different from any other race and because they inherited a secret mark—the swastika and the runic letters SS. Both Liebenfels and List were influential in Vienna between 1908 and 1914, the period when Hitler was living there. During that time, Hitler began to formulate a worldview filled with Haeckel's Darwinian evolution and the racism and anti-Semitism of List and Liebenfels.

These ingredients would later make a witches' brew that would shape his political platform and be eagerly welcomed by many Germans. Perhaps it was prophetic that in the Vienna boarding house where Hitler stayed, it is reported that towards the end of his stay, he sometimes ended his enraged speeches by shouting, "Without Jews, without Rome, We shall build Germany's cathedral! Heil!"[16]

Haeckel popularized Darwinian evolution and pure racism in Germany at the same time that Francis Galton's eugenics found a welcome home in German thought. The German eugenics movement became evident in academic circles in the 1890s, and spread to the public in the early 1900s through journal articles, newspapers, and books. The German populace was very much aware of evolution and its implications.

Galton's eugenics emphasized the role of heredity in predicting physical, mental, and moral traits. The key, then, to the problem of building a better future for Germany was to preserve and advance the Aryan race. With help from Darwin and Haeckel, this racist attitude took on the trappings of science.

Eugenics is very much a natural consequence of

Darwinian human evolution. According to Galton, "The creed of eugenics is founded upon the idea of evolution; not on a passive form of it, but on one that can to some extent direct its own course."[17]

Darwin himself classified humans along racial lines when he determined that black Africans should occupy a lower limb on the tree of life than civilized Europeans. Darwin may not have agreed entirely with everything his favorite disciples said and did with respect to his theory, but it was his ideas that led logically to the conclusions reached in Germany. Eugenics was not an aberration of Darwinism, but a linear consequence of the principles elaborated by Darwin in *Origin*, *Descent*, and *Emotions*.

The drive for German racial purity was given fresh impetus by the Pan-German League, founded in 1890. The goal of this group, of which Haeckel was a founding member, was to unite German-speaking people in one national entity. It adopted anti-Semitism as a policy, because Jews "were unassimilated, dressed distinctively, and remained loyal to traditions."[18] The Pan-German League sought national expansion by annexing lands in Europe and German colonization in Africa. Hitler was an admirer of Pan-German Party leader Georg von Schönerer. He recognized him in *Mein Kampf* for his belief that Jews were an obstacle to German racial purity.

A Horrific Agenda is Born

Hitler's life took on fresh meaning after his World War I service in the German army, where he was wounded and suffered a chlorine gas attack. For his service and sacrifice, he was bemedaled with many military honors, including the Iron Cross.

Instilled with a burning passion for German racial purity, Hitler discovered his gift for oratory in 1919, when he found he had the ability to hold and sway a crowd with his fiery speeches. In 1920 he attracted 2,000 people as he embarked on a political career in the National Socialist German Workers Party (Nazi). His speeches were delivered in guttural tones and employed graphic and crude ideas centered on blaming Jews for the sins of man and exalting the Aryan race. Hitler's message infused hope and pride into many Germans humbled by defeat in World War I and facing the demoralizing terms of the Treaty of Versailles.

During the early 1920s, according to biographer Ian Kershaw, Hitler's "voracious" reading included "influential social-Darwinist and geopolitical tracts…."[19] Hitler overflowed with a deep hatred for Jewish people. He dehumanized Jews by employing terminology that defined them as racial pathogens:

> Don't think that you combat an illness without killing its causative organ, without destroying the

> bacillus, and don't think that you can combat racial tuberculosis without seeing to it that the people is freed from the causative organ of racial tuberculosis. The impact of Jewry will never pass away, and the poisoning of the people will not end, as long as the causal agent, the Jew, is not removed from our midst.[20]

Hitler advanced an agenda rooted in Darwin, Haeckel, Liebenfels, and List. The Jew, without a homeland, was to bear the brunt of his hideous theory built on an evolutionary foundation. According to historian Kershaw,

> A core element—the social Darwinistic view of history as a struggle between individual races with victory going to the strongest, fittest, and most ruthless—seems to have occupied its place at the center of this world-view by 1914-1918 at the latest."[21]

Kershaw's conclusion is widely shared by historians. Richard Weikart explained in the Coral Ridge Ministries video documentary, *Darwin's Deadly Legacy*, that, "Among German historians, there's really not much debate about whether or not Hitler was a social Darwinist; he clearly was drawing on Darwinian ideas."[22] Weikart believes that Hitler, who "was clearly trying to speed evolution along," was not acting alone. "He was drawing on what many other scholars, biologists, and geneticists in Germany were preaching and teaching in the early twentieth century."[23]

Darwin's ideas made Nazi atrocities possible. Anti-Semitism had existed in Europe for centuries before Hitler, but Darwin provided the intellectual catalyst for action. The Nazi program would not have been possible without the ideas of "natural selection and the preservation of favoured races in the struggle for life"—Darwin's subtitle for *Origin*. These ideas did not originate with Hitler. They originated in Darwin's notebook when he drew his tree of life and theorized that good would come from death via natural selection.

My Struggle

Hitler wrote his chilling manifesto, *Mein Kampf*, or *My Struggle*, in 1924-25 while in prison. This summary of his life and political philosophy was a rambling document drafted, translator Ralph Manheim stated, "in the style of a self-educated modern South German with a gift for oratory."[24] The title, *My Struggle*, itself echoes Darwin's "struggle for existence." It is notable that Darwin's German disciple, Ernst Haeckel, also entitled one of his many books, *The Struggle over Evolutionary Thinking* (1905).[25]

Mein Kampf provides overwhelming evidence for the influence Darwin and Haeckel had on Hitler. Here are some examples where we can see without any question how Hitler was influenced by evolutionary principles:

> . . . the purity of the racial blood should be guarded, so that the best types of human beings may be preserved and that thus we should render possible a more noble development [evolution] of humanity itself.[26]

> When man attempts to rebel against the iron logic of Nature, he comes into struggle with the principles to which he himself owes his existence as a man. And this attack must lead to his own doom.[27]

> Thus men, without exception, wander about in the garden of Nature; they imagine that they know practically everything and yet with few exceptions pass blindly by one of the most patent principles of Nature's rule: the inner segregation of the species of all living beings on this earth.[28]

> Here the instinct of knowledge unconsciously obeys the deeper necessity of the preservation of the species, if necessary at the cost of the individual, and protests against the visions of the pacifist windbag who in reality is nothing but a cowardly, though camouflaged, egoist,

> transgressing the laws of development [evolution];...[29]

> The progress of humanity is like climbing an endless ladder: it is impossible to climb higher without first taking lower steps.[30]

> . . . Nature looks on calmly, with satisfaction, in fact. In the struggle for daily bread all those who are weak and sickly or less determined succumb, while the struggle of the males for the female grants the right or opportunity to propagate only to the healthiest. And struggle is always a means for improving a species' health and power of resistance and, therefore, a cause of its higher development [evolution].[31]

> No more than Nature desires the mating of weaker with stronger individuals, even less does she desire the blending of a higher with a lower race, since, if she did, her whole work of higher breeding, over perhaps hundreds of thousands of years, might be ruined with one blow.[32]

Hitler's concepts of nature, race, survival, species, and selection are undeniably connected to Darwin and Haeckel. As scholar Werner Maser has noted, "Darwin was the general source for Hitler's notions in biology, worship, force, and struggle, and of his rejection of moral causality in history."[33]

ENACTING THE DARWINIAN PLAN

Germany was an advanced culture; it was literate, highly cultured and a scientific leader. It has been said that 1939 Germany was the most literate society in the world. Germany's scientific prowess led to heady expectations. Hitler called religion the enemy of science and thought science would doom religion in time. "The best thing," he said, "is to let Christianity die a natural death."[34]

Hitler wanted to harness science to achieve his super race. He would hold forth into the evening, delivering long monologues to Nazi leaders on how the German race should set high standards in science. For Hitler, science would ultimately push the myths of Christianity away. He said,

> The dogma of Christianity gets worn away before the advances of science. Religion will have to make more and more concessions. Gradually the myths crumble. All that's left is to prove that in nature there is no frontier between the organic and the inorganic. When understanding of the universe has become widespread, when the majority of men know that the stars are not

> sources of light but worlds, perhaps inhabited worlds like ours, then the Christian doctrine will be convicted of absurdity.[35]

Hitler and his supporters steadily progressed in numbers and influence from 1920 to 1933. The German people were ready for drastic change in the humiliating aftermath of World War I. The harsh strictures imposed on the nation by the Treaty of Versailles bred discontent and mistrust toward existing governmental leaders. In this climate, Germans became even more receptive to Nazi ideas, which repackaged Darwin, fostered German nationalism, and were openly anti-Semitic.

When Hitler came to power in 1933, he installed a dictatorship with one agenda: enactment of his radical Nazi racial philosophy built on Darwinian evolution. He sought, in Darwin's terms, to preserve the "favoured" race in the struggle for survival. Brute strength and intelligence would be the driving force of the Nazi plan. The first task was to eliminate the weak and those with impure blood who would corrupt the race. These included the disabled, ill, Jews, and Gypsies. Second, the Nazis sought to expand Germany's borders in order to acquire more living space, or "*Lebensraum*," to make room for the expansion of the "favoured" race. Third, the Nazis set about to eliminate communism because of its threat to the Aryan race and because, according to Hitler, communism was the work of

Bolshevik Jews.

The plan quickly unfolded. An order to sterilize some 400,000 Germans was issued within five months of Hitler's rise to power. The order, effective January 1, 1934, listed nine categories of the unfit to be sterilized: feeble-minded, schizophrenia, manic depression, Huntington's chorea, epilepsy, hereditary body deformities, deafness, hereditary blindness, and alcoholism.[36] The Nuremberg Laws were passed in 1935 to prohibit marriage between Jews and Germans and to strip Jews of their German citizenship.

The Nazis established eugenic courts to ensure that the eugenic laws were enforced. To identify the unfit, German eugenicists compared the individual health files of millions of Germans with medical records from hospitals and the National Health Service. The American firm, IBM, aided the effort by automating a national card file system that cross-indexed the defective.[37]

Leading German eugenicist Otmar Freiherr von Verschuer was immersed in the eugenics research effort. Verschuer founded the Institute for Hereditary Biology and Racial Hygiene at Frankfurt University. Its mission, he said, was to be "responsible for ensuring that the care of genes and race, which Germany is leading worldwide, has such a strong basis that it will withstand any attacks from the outside."[38]

Francis Galton, who died in 1911, would likely have been extremely pleased to see his eugenics ideas implemented in Germany. American eugenicists certainly were. They celebrated Verschuer and the German sterilization program. A leading eugenics publication, *Eugenical News*, published an admiring article on Verschuer's Institute that concluded by extending "best wishes to Dr. O. Freiherr von Verschuer for the success of his work in his new and favorable environment."[39]

German and American eugenicists were on excellent terms during the 1930s. Hitler's eugenics program "enjoyed the open approval of leading American eugenicists and their institutions," according to Edwin Black, author of *War Against the Weak*, a study of the American eugenics campaign in the first half of the twentieth century. *The New England Journal of Medicine* editorialized in 1934 that, "Germany is perhaps the most progressive nation in restricting fecundity among the unfit."[40]

American eugenicists believed in the evolution of man and saw eugenics as a means to achieve a more highly evolved America and Germany. Breeding an Aryan race was fully consistent with the ideals of the eugenics movement. After all, once God and His divine authorship of the human race are rejected, moral disorder of every kind quickly becomes possible.

Eugenics in America was not a fringe movement. The U.S. Supreme Court issued a landmark 1927 ruling that authorized the sterilization of a "feeble-minded" Virginia woman. In his majority opinion for the Court, Justice

Oliver Wendell Holmes wrote:

> It is better for all the world, if instead of waiting to execute degenerate offspring for crime, or to let them starve for their imbecility, society can prevent those who are manifestly unfit from continuing their kind. The principle that sustains compulsory vaccination is broad enough to cover cutting the Fallopian tubes. *Jacobson v. Massachusetts,* 197 U.S. 11. Three generations of imbeciles are enough.[41]

After Hitler invaded Poland in 1939, the Nazis became even more aggressive toward the weak. Approximately 100,000 Germans, labeled "useless eaters" by the Nazis, were killed. The victims were patients in nursing homes and medical facilities, as well as Jewish mentally disturbed and disabled. The Nazis ordered all of them euthanized.

This aggressive euthanasia program began by ending the lives of between 70,000 to 80,000 patients in medical and nursing institutions. The Nazis also murdered 10,000 to 20,000 invalids and disabled patients, and some 3,000 children between the ages of three and thirteen who required hospitalization for their special needs.[42] The German churches finally rose up and put pressure on Hitler to stop this very unpopular program. He discontinued it in 1941 and refocused the organization that had been used for this program onto the killing of Jews.

The Third Reich deemed Jews, Gypsies, Jehovah Witnesses, the disabled, homosexuals, Freemasons, and political dissidents as sub-human. The Nazis took away their jobs, their possessions, imprisoned them, and moved them to concentration camps. Ultimately, some 11 million people (and possibly more), of which six million were Jews, were killed by the Nazi death machine. Nearly three million Russian prisoners of war were mostly starved to death, and a half million were executed by Germany.[43]

The most people shot at one time by the Germans was at the Babi Yar ravine in the Ukraine, where the Nazis killed 33,371 mostly Jewish men, women, and children in two days. Until 1942, the chief method of mass execution was to shoot victims beside mass graves. The Nazis began to use closed, unventilated trucks to gas their victims, because it was more effective and less stressful on soldiers. An SS officer stated, "What was uppermost in my mind at that time was that the shootings were a great strain on the men involved and that this strain would be removed by the use of the gas-vans."[44] Not all German soldiers responded the same way. Some took great pleasure in the executions, while others found them repugnant.

Sixteen high-ranking German officials met on January 20, 1942, at a villa overlooking the Grosser Wannsee, a lake outside Berlin, for a conference to organize the "Final

Solution." At this point, most of Europe was under Nazi rule and the Germans were deep into Russia, which seemed almost ready to fall. It was now the time to deal with the Jewish question. The goal was to kill 11 million Jews in Europe.

Were the German soldiers going to shoot them? It was said that it would be demoralizing and ineffective to use German soldiers that way. The discussion invoked Darwinian terms, such as "natural diminution" and "natural selection," indicating the assumption that natural selection was a proven scientific fact.

SS leader Reinhard Heydrich said that the Jews were to be sent east to do forced labor, with the expectation that "a large part of them will be eliminated by natural diminution." Those who survived the harsh labor conditions were to be killed. Using euphemistic language, Heydrich said that they "must be given an appropriate treatment, because they represent a natural selection."[45]

Natural selection implied getting rid of the weak to allow the strongest and most fit to survive. It suggests that as you prune a tree, you can get rid of unwanted branches in the human tree of life. The "final solution" was to be a massive pruning that eliminated millions of so-called weak and bad branches. If man, as Darwin taught, is nothing but an animal, what does it matter? If it is ethical to get rid of unwanted animals, then, under Darwinian logic, it is also ethical to take the lives of 6 million Jews.

The atrocities of Nazism will never be forgotten. The stigma of cruelty, degradation, enraged hate, and the hellish taking of human life will always surround Hitler and the Nazi party. The extreme suffering, torture, pain, and severe humiliation suffered by the Jews and other Nazi victims must not be forgotten. The German people lost their moral and spiritual heritage. Their leaders embraced Darwinian evolution and applied it to questions of race. In so doing they departed from the benevolent Creator and rejected His law.

Is there a link between Darwin and Hitler? Weikart says yes. He outlined, in simplified fashion, the route from Darwin to Hitler:

> First, Darwinism undermined traditional morality and the value of human life. Then, evolutionary progress became the new moral imperative. This aided the advance of eugenics, which was overtly founded on Darwinian principles. Some eugenicists began advocating euthanasia and infanticide for the disabled. On a parallel track, some prominent Darwinists argued that human racial competition and war is part of the Darwinian struggle for existence. Hitler imbibed these social Darwinist ideas, blended in virulent anti-Semitism, and—there you have it: Holocaust.[46]

Today when evolutionists are questioned as to how Darwinian evolution gave birth to Hitler's Nazism, they immediately want to beg the question, answering that racism has nothing to do with science. They are correct! Racism has nothing to do with science, but it has everything to do with evolution—a fact that is unavoidable.

All this can be traced back to the slippery arguments of Darwin himself. He sought to explain in elaborate detail the origin of species, but could not define what a "species" is and was unable to give an adequate definition of "race." He failed dismally in defining these terms. It then gets more slippery when he puts man into the equation and says that man is related to the ape. Placing man on the same plane as the animals makes possible a dangerous application of this bad idea. Man can now be treated like an animal—something grasped and acted upon by the Nazi killing machine.

Adam Sedgwick, a mentor of Darwin's at Cambridge, took immediate issue with *Origin* after reading it in 1859. He told his former student that it "greatly shocked my moral taste" and elaborated in a passage that proved prophetic:

> There is a moral or metaphysical part of nature as well as a physical. A man who denies this is deep in the mire of folly. Tis the crown and glory of organic science that it does thro' final cause, link

> material to moral;... You have ignored this link; and, if I do not mistake your meaning, you have done your best in one or two pregnant cases to break it. Were it possible (which thank God it is not) to break it, humanity in my mind, would suffer a damage that might brutalize it—and sink the human race into a lower grade of degradation than any into which it has fallen since its written records tell us of its history.[47]

The Darwin-driven German Holocaust illustrates what can happen when man forgets and rejects God. When that happens, a vacuum is created into which comes rushing in many false, foolish, and fatal ideas. When He who gave mankind the true Tree of Life is rejected, error and infamy can quickly follow.

CHAPTER EIGHT

FINAL THOUGHTS: FROM DARWIN TO CHRIST

It was dark and cold on those early winter mornings when we left the dorms in silence for the brief walk to the adjoining building. I remember the sense of total peace that enveloped each of us as we kept silence—a mandate for the first two hours of the day. We would quarantine our thoughts within us as we meditated and prayed to the Lord to help us through the day.

All 35 of us adolescent seminarians followed this routine every day, keeping silent for two hours until breakfast. Dressed in black, we would form a natural line, as we moved on the road to the top of the hill, where you could look down at the valley miles away and see a scattering of flickering lonely lights—scant evidence of human activity in the pre-dawn dimness. Every morning we would walk beneath the towering statue of Don Bosco, passing under his gaze as he looked down with outstretched arms. He had two children at his sides as a reminder of our calling to be Catholic priests and brothers who would teach and work with youth and bring them up in love for Christ.

Saint John Bosco founded the Catholic Salesian Society in Turin, Italy, in 1859. It now has over 40,000 priests, brothers, and sisters in over 120 countries. Don (Italian for "Father") Bosco had a heart for poor and needy children and provided technical and vocational training to teach them skills so that they would become productive human beings. The driving force for his mission was the religious training that the boys would receive. Those who were called would join the ranks of the Salesians.

We were junior seminarians—high school age boys attending this seminary with the prospect of taking vows to become priests. It was our daily routine to enter chapel every morning at the same time for mass and prayer. Our schedule kept us busy, as we moved from the chapel to classrooms to study hall and to outdoor sports. We were always actively engaged, with very little idle time. The

Salesians trained us this way to make us capable Salesian teachers, able to instruct others in a manner not ruled by fear or intimidation, but by motivation and encouragement. I never forgot this educational philosophy, one I have applied in my 35 years of teaching. Whether it has been used in seventh grade or at college, I have found that it works.

This was an enjoyable part of my life in the early 1960s. I have fond memories of the small, scenic village of West Haverstraw, which overlooks the Hudson River some 35 miles north of New York City. I spent four years of my life as a dedicated seminarian and was recognized by faculty and students as the mostly likely to succeed as a teaching Salesian brother.

They were wrong.

The summer after my final year at the seminary, I faced the much bigger step of taking the vows for poverty, obedience, and chastity. I had no problem with giving my meager possessions to God's work or with being obedient—I was willing to go wherever they sent me. It was the vow of chastity over which I struggled. This, if taken, would be a pledge to God to never get married. We were taught in the seminary that instead of being involved in the responsibilities that come with marriage, we would marry the Church and become fully committed to its mission.

After much agonizing deliberation, I made my decision to leave, because I knew I wanted to someday get married. I was 18 when I left those beautiful rolling green hills, and I still cherish today all the memories of my years there.

Back home in Brooklyn, I became a commuting college student. Along with tens of thousands of others in the New York City area, I took advantage of inexpensive higher education. I enrolled in college to pursue my passion for science—a love I had acquired at the seminary.

It was in college that I was challenged, even as a chemistry major, with the theory of evolution. One of my first classes was biology, which was taught by an avid evolutionist. I often stayed after class to ask questions and quickly realized that this professor was an atheist who was particularly incensed with Christianity. However, I was not offended at his swipes at my faith. I actually found him quite amusing and convincing.

This young, articulate, and energetic professor was a model teacher who was available to help meet his students' needs. He would speak out against God and announce to his admiring students, a group that included me, that he was an atheist. He would pontificate on the value of being an unbeliever in God, and he was very influential.

Along with this atheist professor, my friends challenged me on my faith in Genesis. They laughed sarcastically when they asked questions like: "How can you believe in Adam and Eve?" "Was there really a Noah's ark or was it a fairy tale?" I began to ponder these questions and the existence of God and wondered, "Do I want to keep my brain in cold storage or use it to face the 'facts' of science?"

I had answered that question even before I read Darwin's *Origin of Species.* I wanted to believe in evolution, but if I believed in evolution, what should I do with the Genesis account and the rest of the Bible? There was no middle ground—no place of compromise for me. I threw out the Bible—and God, too. It was not even six months from my days of silent prayer and meditation at West Haverstraw to my eager embrace of atheism in New York City.

As I look back on my decision, I recognize that I did not critically analyze the claims of evolution. I accepted it because of its emotional appeal as an alternative to God. I read Darwin's *Origin* as an alternative Scripture—something I could place my faith in as reliable and true.

My secular college environment was not conducive to a belief in God. My friends, the biology professor, the class textbook, and the spirit of rebellion in the 1960s all directed me towards freedom from the "establishment." It was disguised as an intellectual decision based on science, but the real truth is that I wanted a drastic change in my life. I wanted to do away with God and His laws, and that is what I did all through my college years and into my teaching career as a science educator. I lost my faith to evolution and became an example of what William

Jennings Bryan said would happen to college students as a consequence of the teaching of evolution.

I not only embraced atheism, but became its eager advocate. God and the Bible angered me now, and it became my agenda to communicate my beliefs when I taught in public school classrooms. I always covered evolution in class, because I thought it was vitally important. I had the ability to teach young, curious minds and found much satisfaction and joy in the classroom teaching experience. Students who enjoyed my science class quickly became friends, which gave me the opportunity to introduce my evolutionary agenda. When they asked questions, my answers always had a way of getting around to atheism and evolution.

Charles Darwin went to Cambridge to study to be a clergyman and was surrounded by the Word of God, just as I was at the seminary. He was mentored by a man at Cambridge who was committed to the Christian faith, but Darwin took another path, and that is exactly what I did. He also gave up his faith and described Christianity in a manner that mimicked my sentiments, calling orthodox belief a "damnable doctrine." This is what I taught, in subtle fashion, in the public school.

The Bible addresses this phenomenon when it says, "The fool has said in his heart, 'There is no God'" (Psalm 53:1). I often wonder how many people really know that the question of faith in God can be reduced to one's heart—the seat of our being—rather than to an intellectual or scientific decision. In an interview, the late Isaac Asimov, one of the most well-known evolutionist intellectuals of his day and the author of hundreds of popular science and science fiction books, said:

> I am an atheist, out and out. It took me a long time to say it. I've been an atheist for years and years, but somehow I felt it was intellectually unrespectable to say one was an atheist, because it assumed knowledge that one didn't have. Somehow it was better to say one was a humanist or an agnostic. I finally decided that I'm a creature of emotion as well as of reason. Emotionally I am an atheist. I don't have the evidence to prove that God doesn't exist, but I so strongly suspect he doesn't that I don't want to waste my time.[1]

Along with Asimov, and I believe Darwin also, I made an emotional choice against God. It all comes down to the heart—where our spirit lives. The choice against God turns not on the evidence, but on emotion—one's sinful rebellion against the Creator.

Evolution is a dangerous idea and one I once held with evangelistic fervor. Now I have a new life in Jesus Christ—a life that began in 1978, after a member of Dr.

Kennedy's church, Coral Ridge Presbyterian, took the time to explain the Gospel to me.

I know what evolution can do. I understand its power to persuade minds eager for an alternative to God. Darwin's theory is an idea that wicked men have used as a rationale for mass murder; it is also an instrument of spiritual death for millions more who embrace its lies. That is why we, as Christians, must do everything we can to refute evolution and to bring the world back to the Creator.

ENDNOTES

FOREWORD

1 Adolf Hitler, *Mein Kampf*, trans. James Murphy, Chapter XI, "Race and People," http://gutenberg.net.au/ebooks02/0200601.txt.
2 Arthur Keith, *Evolution and Ethics*, G.P. Putnam's Sons, New York, 1947, p. 230.
3 Ibid., p.10.
4 Conway Zirkle, *Evolution, Marxian Biology, and the Social Scene*, University of Pennsylvania Press, Philadelphia, 1959. Quoted in Paul G. Humber, M.S., "Stalin's Brutal Faith," *Impact* (#172), Institute for Creation Research, http://www.icr.org/index.php?module=articles&action=view&ID=276.
5 Eduardo del Rio (pseudonym= "Rius"), *Marx for Beginners*, Pantheon Books, New York, 1976, n.p. Quoted in Paul G. Humber, M.S., "Stalin's Brutal Faith."
6 Stephane Courtois, Nicolas Werth, Jean-Louis Panne, Andrzej Paczkowski, Karel Bartosek, Jean-Louis Margolin, *The Black Book of Communism: Crimes, Terror, Repression*, trans. Mark Kramer and Jonathan Murphy, Harvard University Press, Cambridge, Mass., 1999, p. 4. Quoted in D. James Kennedy and Jerry Newcombe, *Lord Of All: Developing A Christian World-and-Life View*, Crossway Books, Wheaton, Ill., 2005, p. 30.
7 "The Big Bang in Astronomy," *New Scientist*, November 19, 1981, pp.521-527. Accessed at http://bevets.com/equotesh2.htm.
8 Jean Rostand, *Age Nouveau* (a French periodical), February 1959, 12, quot ed at http://www.talkorigins.org/faqs/ce/3/part12.html.

CHAPTER ONE

1 Web site of American Museum of Natural History, at http://www.amnh.org/science/papers/odd_sauropod.php.
2 Niles Eldredge, *Darwin: Discovering the Tree of Life*, "Introduction," W. W. Norton & Company, New York/London, 2005, p. 3.

3 Web site of American Museum of Natural History, at http://www.amnh.org/exhibitions/permanent/fossils/.
4 Mark A. Norell, Eugene S. Gaffney, and Lowell Dingus, *Discovering Dinosaurs*, Alfred A. Knopf, Inc., New York, 1995, p.15.
5 Ibid.
6 American Museum of Natural History statement at the entrance to the Darwin Exhibition, New York.
7 Charles Darwin, *On the Origin of Species*, in *The Darwin Compendium*, Barnes & Noble, New York, 2005, p. 447.
8 Ibid., p. 449.
9 Entry for "Species" in *The Stanford Encyclopedia of Philosophy*. Accessed at http://plato.stanford.edu/entries/species/.
10 National Academy of Science, *Teaching About Evolution and the Nature of Science*, National Academy Press, Washington, D.C., 1998, p. 13.
11 Entry for "Species" in *The Stanford Encyclopedia of Philosophy*. Accessed at http://plato.stanford.edu/entries/species/.
12 http://www.thebreedsofdogs.com.
13 Stephen Jay Gould, "Evolution's erratic pace," *Natural History*, Vol. 86, No. 5, pp.12-16, May 1977.
14 John Morris, Ph. D., "Don't the Fossils Prove Evolution?" Institute for Creation Research. Accessed at http://www.icr.org/index.php?module=articles&action=view&ID=525.
15 Ibid.
16 http://www.ucmp.berkeley.edu/history/owen.html.
17 Ibid.
18 Darwin's Exhibition: American Museum of Natural History, New York, 2006.
19 Ibid.
20 Niles Eldredge, pp. 11–12.

CHAPTER TWO

1 Alister McGrath, *The Twilight of Atheism*, Doubleday, New York, 2004, p. 241.
2 Niles Eldredge, *Darwin: Discovering the Tree of Life*, W. W. Norton & Company, Inc., New York/London, 2005, p.19.
3 http://darwin.baruch.cuny.edu/biography/shrewsbury/rdarwin.html.
4 Charles Darwin, *Autobiography*, in *The Darwin Compendium*, Barnes & Noble, New York, p. 1587.
5 E. Janet Browne, *Charles Darwin: Voyaging*, Alfred A. Knopf, New York, 1995, p. 9.

6 Russell Grigg, "Darwinism: it was all in the family," *Creation* magazine, Vol. 26, December 2003, (1): 16-18. Accessed at http://www.answersingenesis.org/creation/v26/i1/darwinism.asp.
7 Darwin, *Autobiography*, in *The Darwin Compendium*, p. 1585.
8 Plaque in memory of Charles Darwin. Accessed at: http://www.baruch.cuny.edu/wsas/darwin/biography/shrewsbury/unitarian.html.
9 F. Darwin, ed., *The Life and Letters of Charles Darwin*, New York, D. Appleton & Co., 1905, 348. Accessed at http://pages.britishlibrary.net/charles.darwin/texts/letters/letters1_09.html.
10 Darwin, *Autobiography*, in *The Darwin Compendium*, p. 1587.
11 Ibid.
12 Ibid.
13 Information on The Lunar Society. Accessed at http://jquarter.members.beeb.net/morelunar.htm.
14 Erasmus Darwin, *Zoonomia, or the Laws of organic life*, Thomas and Andrews, 2nd American Edition, Vol. 1, Boston, p. 397.
15 http://www.answersingenesis.org/creation/v26/i1/darwinism.asp#f13.
16 http://www.ucmp.berkeley.edu/history/Edarwin.html.
17 Darwin, *Autobiography*, in *The Darwin Compendium*, p. 1590.
18 William Jennings Bryan, "Mr. Bryan on Evolution," *Readers' Digest*, vol. 4, No. 40. Accessed at http://bevets.com/equotesb.htm.
19 Grigg, *Creation* magazine, Vol. 26, December 2003, (1): 16-18.
20 D. King-Hele, *Erasmus Darwin*, Charles Scribner's Sons, New York, 1963, p. 70. Quoted by Grigg, "Darwinism: it was all in the family," *Creation* magazine, Vol. 26, December 2003, (1): 16-18.
21 Grigg, "Darwinism: it was all in the family," *Creation* magazine, Vol. 26, December 2003, (1): 16-18.
22 Darwin, *Autobiography*, in *The Darwin Compendium*, p. 1590.
23 Ibid.
24 Ibid.
25 http://en.wikipedia.org/wiki/Robert_Edmond_Grant
26 Darwin, *Autobiography*, in *The Darwin Compendium*, p. 1593.
27 Ibid.
28 F. Darwin, ed., *The Life and Letters of Charles Darwin*, p. 39. charles.darwin/texts/letters/letters1_02.html.
29 Ibid., p. 40.
30 Darwin, *Autobiography*, in *The Darwin Compendium*, p. 1594.

CHAPTER THREE

1 Charles Darwin, *Autobiography*, in *The Darwin Compendium*, Barnes & Noble, New York, 2005, p. 1596.
2 S.M. Walters and E.A. Stow, *Darwin's Mentor*, University of Cambridge, Cambridge, 2001, p. 79.
3 Darwin, *Autobiography,* in *The Darwin Compendium*, p. 1596.
4 Walters and Stow, *Darwin's Mentor*, p. 160.
5 Ibid., p. 23.
6 http://www.ucmp.berkeley.edu/history/sedgwick.html.
7 Darwin, *Autobiography*, in *The Darwin Compendium*, p. 1594.
8 See, for example, Genesis 1:24: "Then God said, 'Let the earth bring forth the living creature according to its kind: cattle and creeping thing and beast of the earth, each according to its kind'; and it was so" (NKJV). Also see Genesis 1: 11-12, 21.
9 Henry Morris, *The Long War Against God*, Master Books, Green Forest, Arkansas, 1989, p. 189.
10 Terry Mortensen, *The Great Turning Point*, Master Books, Green Forest, Arkansas, 2004, p. 30.
11 H.H. Read, *The Granite Controversy*, 1957, p. xi., Quoted in Terry Mortenson, *The Great Turning Point*, Master Books, Green Forest, Arkansas, 2004, p. 11.
12 Darwin, *Autobiography*, in *The Darwin Compendium*, p. 1600.
13 Walters and Stow, *Darwin's Mentor*, p. 95.
14 Ibid., p. 34.
15 Stephen J. Gould, "Pillars of Wisdom," *Natural History*, April 1999.
16 Morris, *The Long War Against God*, p. 189.
17 http://www.fossilnews.com/2000/grnrv/grnrv.html.
18 Ibid., p. iv.
19 Darwin, *Autobiography*, in *The Darwin Compendium*, p. 1601.
20 Walters and Stow, *Darwin's Mentor*, p. 95.
21 Morris, *The Long War Against God*, 1989, p.167.
22 Darwin, *Autobiography*, in *The Darwin Compendium*, p. 1708.
23 http://pages.britishlibrary.net/charles.darwin2/texts.html.
24 Darwin, *Autobiography*, in *The Darwin Compendium*, p.1687.
25 Niles Eldredge, *Darwin: Discovering the Tree of Life*, W.W. Norton & Company, New York/London, 2005, p. 45.
26 Walters and Stow, *Darwin's Mentor*, p. 169.
27 Ibid., p. 170.
28 Ibid., p. 173.
29 Darwin, *Autobiography* in *The Darwin Compendium*, p. 1603.

30 F. Darwin, ed., *The Life and Letters of Charles Darwin*, D. Appleton & Co., New York, 1905. p. 43. Accessed at http://pages.britishlibrary.net/charles.darwin/texts/letters/letters2_01.html.
31 Walters and Stow, *Darwin's Mentor*, p.173.
32 F. Darwin, ed., *The Life and Letters of Charles Darwin*, p. 157. Accessed at http://pages.britishlibrary.net/charles.darwin/texts/letters/letters1_04.html.

CHAPTER FOUR

1 Charles Darwin, *Autobiography*, in *The Darwin Compendium*, Barnes & Noble, New York, 2005, p. 1610.
2 Ibid.
3 http://www.guinnessworldrecords.com/content_pages/record.asp?recordid=48417.
4 Darwin, *Autobiography*, in *The Darwin Compendium*, p. 1639.
5 Ibid, p. 1687.
6 Darwin, *On the Origin of Species*, 6th ed. 1872, p. 1. Accessed at http://pages.britishlibrary.net/charles.darwin/texts/origin_6th/origin6th_fm.html.
7 Niles Eldredge, *Darwin: Discovering the Tree of Life*, W. W. Norton & Company, New York, 2005, p. 89.
8 Carl Wieland, "Darwin's Finches," *Creation* magazine, 14 (3):22-23.
9 Ibid.
10 Eldredge, *Darwin: Discovering the Tree of Life*, p. 63.
11 http://www.rit.edu/~rhrsbi/GalapagosPages/DarwinFinch.html.
12 Ibid.
13 Eldredge, *Darwin: Discovering the Tree of Life*, p. 103.
14 Ibid., p. 108.
15 Ibid.
16 Ibid., p. 66.
17 Thomas Malthus, *An Essay on the Principle of Population*, Chapter 1. Accessed at http://www.ac.wwu.edu/~stephan/malthus/malthus.1.html.
18 Thomas Malthus, *An Essay on the Principle of Population: A View of its Past and Present Effects on Human Happiness; with an Inquiry into Our Prospects Respecting the Future Removal or Mitigation of the Evils which It Occasions*, John Murray, London, 1826, Sixth edition, Footnote 7. Accessed at http://www.econlib.org/library/Malthus/malPlong.html.
19 Charles Darwin, *The Autobiography of Charles Darwin* (paperback edition), W. W. Norton & Company, New York/London, 1958, p. 120.
20 Darwin, *On the Origin of Species*, in the *The Darwin Conpendium*,

Barnes & Noble, New York, 2005, pp. 469–470.

21 Ibid., p. 471.

22 Ibid. p. 1726.

23 Darwin, *The Autobiography of Charles Darwin*, p. 59.

24 William Paley, *Natural Theology; or, Evidences of the Existence and Attributes of the Deity*, J. Faulder, London, 1809, pp. 542-543. Republished electronically by University of Michigan Humanities Text Initiative. Accessed at http://www.hti.umich.edu/cgi/p/pd-modeng/pd-modeng-idx?type=HTML&rgn=TEI.2&byte=53049319.

25 Nora Barlow, ed., *The Autobiography of Charles Darwin, 1809-1882*: with original omissions restored, W. W. Norton & Company, New York, 1969, p. 87.

26 F. Darwin, ed., *The Life and Letters of Charles Darwin*, D. Appleton & Co., New York, 1905, p. 274. Accessed at http://pages.britishlibrary.net/charles.darwin/texts/letters/letters1_08.html. (Addressed to Mr. J. Fordyce, and published by him in his "Aspects of Scepticism," 1883.)

27 Ibid., p. 529.

28 Ibid, pp. 91, 92.

29 Barlow, ed., *The Autobiography of Charles Darwin, 1809-1882*: with original omissions restored, p. 85.

30 Ibid., p. 85.

31 Ibid., pp. 86-87.

32 Ibid., p. 236.

33 Ibid., p. 238.

CHAPTER FIVE

1 Charles Darwin, *On the Origin of Species*, John Murray, London, 1859, p. 488. Accessed at http://pages.britishlibrary.net/charles.darwin/texts/descent/descent01.html.

2 F. Darwin, ed., *The Life and Letters of Charles Darwin*, D. Appleton & Co., New York, 1905, p. 76. Accessed at http://pages.britishlibrary.net/charles.darwin/texts/letters/letters1_02.html.

3 Jennifer Harper, "Americans still hold faith in divine creation," *The Washington Times*, June 9, 2006. Accessed at http://washingtontimes.com/national/20060608-111826-4947r.htm.

4 Janet Browne, *Charles Darwin: The Power of Place*, Princeton University Press, Princeton, New Jersey, 2002, p. 362.

5 Mark P. Cosgrove, *The Amazing Body Human: God's Design for Personhood*, Baker Book House, Grand Rapids, Michigan, 2002, p. 33.

6 Ibid., pp. 30-31.
7 Charles Darwin, *The Descent of Man and Selection in Relation to Sex*. 2nd ed., John Murray, London, 1882, p. 2. Accessed at http://pages.britishlibrary.net/charles.darwin/texts/descent/descent_front.html.
8 Werner Gitt, *The Wonder of Man*, Christliche Literatur-Vertritung e.v., Bielelefid, Germany, 1999, p. 81.
9 Ibid.
10 Browne, *Charles Darwin: The Power of Place*, p. 343.
11 Darwin, *The Descent of Man,* Preface in *The Darwin Compendium*, p. 732.
12 Darwin, *On the Origin of Species*, 6th ed., 1872, p. 449. Accessed at http://pages.britishlibrary.net/charles.darwin/texts/origin1859/origin13.html. Quoted in Jonathan Wells, *Icons of Evolution*, Regnery Publishing, Washington, D.C., 2000, p. 81.
13 Darwin, *On the Origin of Species*, 6th ed., p. 395.
14 F. Darwin, ed., *The Life and Letters of Charles Darwin*, D. Appleton & Co., New York, 1905, p. 131. Accessed at http://pages.britishlibrary.net/charles.darwin/texts/letters/letters2_02.html. Quoted in Jonathan Wells, *Icons of Evolution,* Regnery Publishing, Washington, D.C., 2000, p. 82.
15 Darwin, *Descent,* 2nd ed., p. 3.
16 Ibid., p. 205.
17 Darwin, *Origin*, p. 33.
18 Darwin, *Descent,* 2nd ed., p. 146.
19 Ibid., pp. 176-177.
20 Ibid., p. 180.
21 Ibid., p. 167.
22 "The Proper Study of Mankind," *The Economist*, July 1, 2000, p. 11. Accessed at http://www.as.ua.edu/ant/bindon/ant275/reader/The%20proper%20study%20of%20mankind.pdf.
23 Luigi Luca Cavalli-Sforza & Francesco Cavalli-Sforza, *The Great Human Diasporas: The History of Diversity and Evolution,* Addison-Wesley, New York, 1995, p. 124. Accessed at http://www.assemblage.group.shef.ac.uk/2/2evison2.html.
24 Browne, *Charles Darwin: The Power of Place*, p. 362.
25 Darwin, *Descent,* 2nd ed., p. 183.

CHAPTER SIX

1 Adrian Desmond and James Moore, *Darwin: The Life of a Tormented Evolutionist,* W. W. Norton & Company, New York/London, 1991, p. 476.
2 Janet Browne, *Charles Darwin: The Power of Place*, Princeton University

Press, Princeton, New Jersey, 2002, p. 407.

3 Desmond and Moore, *Darwin: The Life of a Tormented Evolutionist*, p. 249.

4 Ibid., pp. 476-77.

5 Nora Barlow, ed., *The Autobiography of Charles Darwin, 1809-1882:* with original omissions restored, W. W. Norton & Company, New York/London, 1969, p. 106.

6 Letter from Thomas Huxley to Charles Darwin, November 23, 1859. Accessed at http://aleph0.clarku.edu/huxley/letters/59.html#23nov1859.

7 Leonard Huxley, *Life and Letters of Thomas Henry Huxley*, 2 vols., London, 1900, Vol. I, 179. Quoted in J. R. Lucas, "Wilberforce and Huxley: A Legendary Encounter." Accessed at http://users.ox.ac.uk/~jrlucas/legend.html#r-6.

8 Browne, *Charles Darwin: The Power of Place*, p. 221.

9 Thomas Huxley, *Lay Sermons, Addresses and Reviews*,Appleton, New York, 1871, p. 20. Quoted in Henry Morris, "Evolution and Modern Racism (#7)," Impact, Accessed at http://www.icr.org/index.php?module=articles&action=view&ID=55. Emphasis added.

10 Desmond and Moore, *Darwin: The Life of a Tormented Evolutionist*, p. 532.

11 Ibid.

12 http://www.galton.org/books/memories/chapter-XX.html.

13 Desmond and Moore, *Darwin: The Life of a Tormented Evolutionist*, p. 665.

14 Edwin Black, *War Against the Weak*, Four Walls Eight Windows, New York, 2003, p. 14.

15 Ibid., p. 15.

16 Francis Galton, *Hereditary Genius: An Inquiry Into Its Laws and Consequences*, Macmillan & Co., London, 1892, p. 31. Accessed at http://www.galton.org/books/hereditary-genius/text/html/galton-1869-genius.html.

17 Letter from Darwin to Galton. Accessed at http://galton.org/letters/darwin/correspondence.html.

18 Niles Eldredge, *Darwin: Discovering the Tree of Life*, W. W. Norton & Company, New York/London, 2005, p. 13.

19 Francis Galton, *Essays in Eugenics*, University Press of the Pacific, Honolulu, Hawaii, 2004, p. 35.

20 Ibid., p. 24.

21 Francis Galton, "Eugenics: Its Definition, Scope, and Aims," *The American Journal of Sociology*, Vol. X, No. 1, July 1904. Quoted in Black, *War*, p. 18.

22 Margaret Sanger, *The Pivot of Civilization*, Brentano's, New York, 1922. p. 189. Quoted in Black, *War*, p. 129.

23 Black, *War Against the Weak*, p. 82. Emphasis added.

24 Browne, *Charles Darwin: The Power of Place*, p. 250.
25 Galton, *Essays in Eugenics*, pp. 68-69.
26 Desmond and Moore, *Darwin: The Life of a Tormented Evolutionist,* p. 477.
27 Alister McGrath, *The Twilight of Atheism*, Doubleday, New York, p. 83.
28 Russell Grigg, "Evangelist for evolution and apostle of deceit," *Creation* 18(2):33–36, March 1996. Accessed at http://www.answersingenesis.org/creation/v18/i2/haeckel.asp#f15.
29 Ernst Haeckel entry in 1911 Encyclopedia. Accessed at http://www.1911encyclopedia.org/Ernst_Heinrich_Haeckel.
30 Desmond and Moore, *Darwin*, p. 540.
31 Charles Darwin, *Autobiography of Charles Darwin,* from *The Darwin Compendium,* Barnes & Noble, New York, 2005, p. 1780.
32 Desmond and Moore, *Darwin*, p. 539.
33 Ibid., p. 538.
34 Ibid., p. 539.
35 Ibid., p. 543.
36 Ernst Haeckel, *Eric Weisstein's World of Biography*. Accessed at http://scienceworld.wolfram.com/biography/Haekel.html.
37 Grigg, "Evangelist for evolution and apostle of deceit," *Creation* 18(2):33–36, March 1996.
38 Charles Darwin, *The descent of man and selection in relation to sex*. 2nd edn., John Murray, London, 1882, p. 3. Accessed at http://pages.britishlibrary.net/charles.darwin/texts/descent/descent_front.html.
39 Darwin, *Autobiography*, in *The Darwin Compendium*, p. 1781.
40 Ibid., p. 1782.
41 Thomas Huxley, *Critiques and Addresses* (1873), Project Gutenberg, released June 3, 2004. Accessed at http://www.gutenberg.org/files/12506/12506-8.txt. Emphasis added.
42 Desmond and Moore, *Darwin*, p. 561.
43 Grigg, "Evangelist for evolution and apostle of deceit," *Creation* 18(2):33–36, March 1996.
44 Desmond and Moore, *Darwin: The Life of a Tormented Evolutionist,* p. 543.
45 Quoted in Richard Weikart, *From Darwin to Hitler*, Palgrave MacMillan, New York, 2006, p. 106.
46 Weikart, *From Darwin to Hitler*, p. 110.
47 Ibid., p. 147.
48 Ibid.
49 Ibid., p. 148.
50 Ibid., p. 187.

51 Ibid.

CHAPTER SEVEN

1 Janet Browne, *Charles Darwin: The Power of Place*, Princeton University Press, Princeton, New Jersey, 2002, p. 141.
2 Alister McGrath, *The Twilight of Atheism*, Doubleday, New York, 2003, p. 59.
3 Richard Weikart, *From Darwin to Hitler,* Palgrave MacMillan, New York, 2006, p. 187.
4 Ibid.
5 Ibid., p. 188.
6 Ibid., p. 26.
7 Ibid., p. 29.
8 Ibid.
9 Ibid., p. 30.
10 Ibid., pp. 39-40.
11 Ibid., p. 195.
12 Robert Payne, *The Life & Death of Adolf Hitler*, Barnes & Noble, New York, 1995, p. 58.
13 Ibid., p. 57.
14 Robert Waite, "The Intellectual Roots of Hitler's Anti-Semitism," in Brenda Stalcup (ed.), *People Who Made History: Adolf Hitler*, Greenhaven Press, San Diego, 2000, p. 43.
15 Ibid., p. 44.
16 Payne, *The Life & Death of Adolf Hitler*, p. 91.
17 Francis Galton, *Essays in Eugenics*, University Press of the Pacific, Honolulu, 2004, p. 68.
18 Dick Geary, "Anti-Semitism in Austria," in Brenda Stalcup (ed.), *People Who Make History: Adolf Hitler*, Greenhaven Press, San Diego, 2000, p. 43.
19 Ian Kershaw, *Hitler*, Pearson Education, Harlow, England, 2001, p. 26.
20 Ibid., pp. 27-28.
21 Ibid., p. 26.
22 *Darwin's Deadly Legacy*, Coral Ridge Ministries, 2006. To obtain this resource, please go to www.coralridge.org or call 1-800-988-7884.
23 Ibid.
24 Adolf Hitler, *Mein Kampf* , trans. by Ralph Manheim, "Translator's Note," Houghton Mifflin Company, Boston/New York, 1943, p. ix.
25 Paul G. Humber, "The Ascent of Racism," *Impact*, (#164) Institute for Creation Research, El Cajon, California, February, 1987. Accessed at http://www.icr.org/index.php?module=articles&action=view&ID=268.

26 Adolf Hitler, *Mein Kampf*, trans. by James Murphy. Accessed at http://gutenberg.net.au/ebooks02/0200601.txt.
27 Adolf Hitler, *Mein Kampf*, Houghton Mifflin Company, Boston, 1943, p. 287.
28 Ibid., p. 284.
29 Ibid., p. 299.
30 Ibid., p. 295.
31 Ibid., p. 285.
32 Ibid., p. 286.
33 Ian Taylor, *In the Minds of Men*, TTFE Publishing, Toronto, 1987, p. 409.
34 John Cornwell, *Hitler's Scientists*, Penguin Books, New York, 2003, p. 35.
35 Ibid., pp. 35-36.
36 Edwin Black, *War Against the Weak*, Four Walls Eight Windows Publishers, New York, 2003, p. 299.
37 Ibid., p. 339.
38 Ibid., p. 340.
39 Ibid., pp. 341-42.
40 David Morgan, "Yale Study: U.S. Eugenics Paralleled Nazi Germany," *Chicago Tribune*, February 15, 2000. Accessed at http://www.commondreams.org/headlines/021500-02.htm.
41 *Buck v. Bell*, 274 U.S. 200 (1927). Accessed at http://www.law.du.edu/russell/lh/alh/docs/buckvbell.html.
42 Sebastian Haffner, "The Holocaust Cost Hitler Victory in the War," in Brenda Stalcup (ed.), *People Who Made History: Adolf Hitler*, Greenhaven Press, San Diego, 2000, p. 112.
43 Ibid., p. 115.
44 Ernst Klee, Willi Dressen, and Volker Riess, Editors, *The Good Old Days*, Konecky & Konecky, Old Saybrook, Conn., 1988, p. 69.
45 Payne, *The Life & Death of Adolf Hitler*, p. 466.
46 Weikart, *From Darwin to Hitler*, p. 3.
47 Payne, *The Life & Death of Adolf Hitler*, p. 1.

CHAPTER EIGHT

1 Isaac Asimov, in "An Interview with Isaac Asimov on Science and the Bible," *Free Inquiry* magazine, Spring 1982.

INDEX